Superhouse

Superhouse

Don Metz

GARDEN WAY PUBLISHING ❦ CHARLOTTE, VERMONT 05445

Illustrations by Ed Epstein

Library of Congress Cataloging in Publication Data
Metz, Don.
  Superhouse

  Includes index.
  1. Solar houses. 2. Solar energy—Passive systems. 3. Dwellings—Energy conservation. I. Title.
TH7414.M47   690'.869   81-13251
ISBN 0-88266-258-9   AACR2

*This book is dedicated to the contractors and carpenters
who have shown me the wisdom and beauty of common sense,
especially to Bill Porter, Leonard Cook, Wayne Pike,
Eddie Pierson, and Tom Proe.*

CONTENTS

R O O T S

I grew up on a sixty-acre farm in the rolling hills of southeastern Pennsylvania. My father was a businessman and spent most of his time away at work, but his heart was always there on the farm. He and my mother spent their weekends "fixing it up," not so much because it needed fixing, but because they so enjoyed doing it. My earliest and fondest memories are of projects we would do together around the place—fencing, building, rebuilding, planting, clearing brush and haying. It was absorbing and natural, and I've never thought of the work involved as anything other than fun.

Our tillable acreage was leased to a neighbor who planted cash crops. Farther up the hill, beyond the furrowed rows of corn and soybeans, was a hayfield, waist-high in timothy. Twice each summer we'd crank up the old Ford tractor and Ferguson baler, replenish the grease and oil, and make enough hay to feed the horses over the winter.

From the top of the hayfield, looking south, we could see the distant corners of the farm. The house lay centered in the property, at the end of a long lane that came in straight from the town road. To the east of the house was a springhouse, and to the east of that, a pond. To the south of the house, about 500 feet away, were the barn, corncrib, and silo. The buildings were built primarily of stone, laid up nearly 200 years ago by German settlers. To the west was a hillside orchard, apples, plums, and pears, where horses grazed and waited for the ripe fruit to drop to the ground between the ancient trunks. South of

The springhouse.

A reclusive frog and a couple of crayfish lived there year round, with embarrassingly few places to hide in the crystal-clear water.

2

Sodas, frogs, and crayfish.

the barn, open pasture extended downslope to the property line, where Osage orange, mulberry, and sassafras grew thick with rusty barbed wire, long unneeded, woven in between.

Dependable Spring

At the southeast corner of the hayfield was a huge silver beech tree, and at the base of the tree, bubbling up miraculously from between its roots, was a year-round spring, cold and clear. From this source began a watershed which flowed, appearing and disappearing, down a dark green crease to our springhouse. The springhouse was a small two-story building, whitewashed fieldstone, with a cedar shingled roof. The upper story was a single room with a huge stone fireplace, three windows and a low-posted door. The lower level was a full story below grade on three sides. The fourth side allowed access down a flight of stone steps into a tiny courtyard protected by a shed roof that came off the gable end of the building above. From there, a door opened into the springhouse proper.

A stone peninsula led from the door out into the room. The peninsula was surrounded by a moat, three feet across, a foot and a half deep. A reclusive frog and a couple of crayfish lived there year round, with embarrassingly few places to hide in the crystal-clear water. A big galvanized pressure tank and a squat shallow well pump sat up on a stone base off to one corner. A length of iron pipe with a foot valve on the end extended out into the water about six inches off the bottom.

It was serene and special in that room and it was comforting to know that the water for our house and barn came from there, sweet, soft, and never-failing.

The overflow from the springhouse, as it made its way through watercress and skunk cabbage down to the pond, would freeze up quickly as December's sun dropped farther towards the horizon.

But even in below-zero weather the water inside the springhouse would never freeze. Buried almost eight feet deep on three sides, the thick stone walls and floor were in direct contact with the moderating effects of the earth. Except for the frozen top foot or so, the earth around it maintained a temperature 15° to 20° F. above freezing. Added to that influence was the constant incoming flow of water, its temperature determined by the earth itself. Because of the quantities of earth, stone, and water involved, and because they were all geothermally tempered, a 100° winter-summer variation in outside temperatures would change the inside springhouse temperature only about 10° F. either side of its 45° F. average.

Always Cool

In the summer, the springhouse was always cool. We'd sink watermelon and R.C. Cola into the chilly water to reclaim during hot afternoons. The previous owners stored their milk and homemade butter there, on oak

shelves still standing solid but mildewed near the pump. In the winter, after a cold day's skating on the pond, we'd occasionally clump down the stone steps and open the door to the spring room. Vapor would waft out at us from the earth- and water-warmed room. If we stayed a while, searching in the dark water for the hibernated frog or gulping a long-forgotten soda pop, the snow from our boots would melt on the stone floor.

Water to House

From the springhouse, a lead pipe carried the water underground to the house. From the house, another pipe supplied the barn. Where the pipe crossed under the lane, it was buried a foot and a half deeper than elsewhere to avoid freezing when the lane was plowed clear of snow. Even if the ground was frozen stiff before the first snowfall, it would thaw itself out by February. An insulating blanket of snow works so well that the earth's temperature is sufficient to melt the frost from below.

Effective Insulation

As insulation, snow seems an odd material, and yet at low temperature ranges it's quite effective. Trapped air space is the critical element in any insulation. Our conventional high-tech insulating materials are not only full of trapped air spaces, but they meet other important criteria as well. They're chemically inert, inorganic (inedible), fireproof, and can be shaped into uniform dimensions. And they don't melt at 33° F.

The thousands of cavities between flakes in a block of snow justly qualify it as an insulator. After a big snowstorm Dewey Thompson's snowplow would leave a mountain of snow at the north end of the dooryard. Within hours we kids would have hollowed out our version of an igloo, with a piece of pond ice for a window.

Stuffed with three or four red-cheeked, wet-mittened friends, a kerosene lamp, and a long-haired dog, the igloo seemed improbably warm. A burlap bag kept the wind out and, after a while, the walls would glaze over as the snow melted and turned to ice. The kerosene lamp and body heat in that small enclosure provided the warmth, although to be sure, the thermal comfort we sensed was only relative. The 40° F. temperatures we generated inside the igloo *were* moderate compared to the icy wind which blew down the hill through the orchard and across our hidden winter palace.

My three sisters were horse enthusiasts and, whether by conspiracy or not, I, as the only boy, seemed destined to see more of the dung fork than the bridle. But I admit I liked working in the barn. It was a job reserved for the winter months, after school, after dark. There were horses to feed and four box stalls to keep clean. It was quiet and snug in the barn, fragrant with timothy hay dust, straw bedding, and animal smells. The barn was a typical Pennsylvania Dutch sidehill structure—huge—with massive stone ends and a

high gable roof. The three-story haymow at the upper level was framed out with mortised and tenoned oak sills and plates, posts and beams, knee braces, cross ties, rafters, and purlins. Below was a series of stalls and stanchions, arranged two-deep along the buried back wall. These opened out at grade on the downhill side with the haymow projecting out from above to form an arcade the length of the barn.

Light Switch

There was a light switch on the wall where the path from the house met the corner of the barn. With a click, a long row of cobwebbed bulbs brought the enormous building

Hay insulation and earth-berming.

over the world. Primitive? Sure. But the underlying principle is applicable to many contemporary building types as we will see in the following chapters.

The walls of the old farmhouse were stone, sixteen inches thick, plastered and painted on the inside, stuccoed and whitewashed on the outside. The original builders left the walls uninsulated—not from lack of care, but because there were no appropriate materials available. To keep warm, we put in a hot water baseboard heating system and opened up the three old fireplaces. As was customary in many pre-OPEC homes, the unending drone of the oil burner in the basement was an accepted part of the acoustical scenery. The amounts of oil consumed were oceanic. Primarily at fault were those two-story high stone walls. Stone is a good conductor of heat, which means it's also a poor insulator. (No trapped air spaces). The insulating value of a foot and a half of stone is roughly equivalent to the insulating value of an inch and a half of wood. Plus, its specific heat is much lower than wood, so it always *feels* colder and is slower to change temperature from cold to hot. The old house was solid, fire resistant, and cool in the summer, but it would have been better insulated if its walls had been replaced by nothing but double-paned glass.

Few Windows

As if they were meant to save heat, the windows were small and infrequent. On the

to life, and the horses would grunt and stomp in anticipation of sweet-feed and attention. First the water buckets were collected and refilled. The water in the back stalls never froze. Even the front stalls stayed above freezing most of the time. There was no heat put into the barn other than the body heat the horses generated. A fifteen-foot depth of hay on the floor above the low-ceilinged stalls provided maximum insulation against heat loss. As in the springhouse, the earth-coupled back wall and floor surrounded the space with low-grade but still above-freezing temperatures. These kept the animals comfortable and kept ice out of their water buckets. The design concept of that barn is found in various forms in temperate climate cultures all

Bandit, the passive solar connoisseur.

south side of the house, a big sycamore shaded the building in summer. The windows to the south were larger and more plentiful than on any other wall. When the leaves fell and scraps of the brown bark peeled off the patchy white trunk, the lowered winter sun beamed in under the branches and warmed the summer-dark interior corners of the rooms.

The most appreciative recipient of this annual solar bonus was Bandit, our family dog. Bandit was the foremost active beneficiary of passive solar heating of his time. His sense of east-west was infallible. All day long, like clockwork, he anticipated the radiant patches of light as they infiltrated the house, tracking them from room to room. He knew a warm floor when he felt one. And the words *passive solar* were never once uttered in his presence.

Back in the sixties, *modular* was a one-size-fits-all term prefixed to all sorts of building

trade components. It seems that passive solar have become the pop-tech buzzwords of the Solar Age. The word modular was demystified when the public finally realized that even a common brick was modular, and the word lost its whammy. *Passive solar* may also find its place in the verbal junkyard when enough domesticated animals make it clear that the term simply refers to sunlight through a window. Meanwhile, we can expect to see the word *solar* incongruously coupled with philosophies and products the sun-worshipping Egyptians never dreamed of. Solar burgers, solar politics, solar Zen, and solar disco are in the making at this very moment.

The dog, of course, was right. Sunlight is warm. And, so far, sunlight is free. During the winter months, Philadelphia receives 990 Btu per square foot per day. Buffalo receives 540, Boise 1,090, and Portland, Oregon, 570. Tucson gets 1,800 and Wichita gets 1,410. A

Btu (British thermal unit) is a measure of the amount of energy (heat) required to raise one pound of water 1° F. A modern airtight wood parlor stove produces around 40,000 Btu per *hour*. The typical American kitchen stove with all its burners turned on and its oven up to 500° produces a similar amount. A pilot light on a gas cookstove produces about 75 Btu each hour. People sitting in a movie theater put out 200–300 Btu per hour, even in a movie rated P.G. Refrigerators, light bulbs, and TV sets give off thousands of Btu every day as a by-product of their primary function. When we talk about energy-efficient buildings we use the Btu as the most convenient yardstick. How many Btu does the building require to offset its loss of heat through windows, walls, and roof?

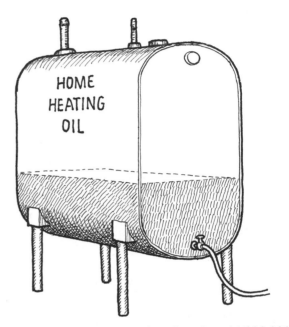

You fill your 100-gallon oil tank and get 14,000,000 Btu.

High Heat Loss

On the coldest night of the year the heat loss from our farmhouse may have approached 100,000 Btu each hour. There are 140,000 Btu in a gallon of no. 2 heating oil. In a well-tuned burner, only 70 to 75 percent of this energy is convertible to usable heat.

$$\frac{140,000}{100,000} \times 70\% = .98 \text{ gallon per hour.}$$

Multiplied by current oil prices, an ongoing daily heat loss in the range of 100,000 Btu/hr is unacceptable.

If we had glazed an area ten feet by thirty feet on the south side of the house, we would have had the opportunity to pick up (300

Most of my education was a matter of language, learning to describe the natural phenomena that I grew up with on the farm.

8

square feet × 990 Btu) an average of 297,-000 Btu per day during the winter.

Over twenty four hours, in the worst of weather, let's assume that old house lost an average of 80,000 Btu per hour. So 80,000 × 24 hours = 1,920,000 Btu per day offset by 297,000 gain through the south glazing. The solar gain accounts for only 15 percent of the Btu required to offset the heat loss in this example. But consider those cold stone walls. The same house insulated to today's standards would require in the neighborhood of 20,000 Btu per hour, a fourth of its original consumption. This puts the solar contribution up to 60 percent and makes a low-tech greenhouse begin to look attractive.

When I went away to college and "learned" about building and engineering, most of my education was a matter of lan-

guage, learning to describe the natural phenomena that I grew up with on the farm. Back home, it was always understood, for instance, that after a hard morning's ride in cool weather, the horses would be blanketed and walked around for half an hour so they'd cool off slowly, and thus the perspiration wouldn't leave their bodies too quickly. Without knowing the language, we were slowing the effect of *evaporative cooling* and preventing a chill.

Driving up the town road in winter, over the frozen brook, we found the bridge surface froze long before the roadbed on either side of it. Years later, on a trip back home, I noticed a sign to that effect—"Caution, bridge freezes first"—telling everyone what most of us already knew. The *earth-coupled* roadbed didn't freeze because it was *geothermally mod-*

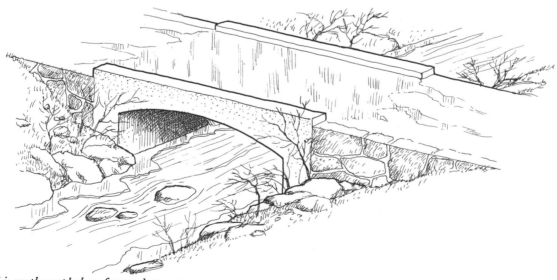

Roadbed is earth-coupled, so freezes last.

erated. The bridge froze because it was exposed to the cold air on all sides and assumed the *ambient air temperature.* As kids, building secret hideouts in the underbrush, we learned that scraping the ground clear of leaves and grass left us with an earthern floor that was cold to lie on. A carpet of dry pine needles and leaves made a warm bed because we were *insulated* from the colder ground. We took advantage of the difference in *specific heat* between the wet earth and the dry plant material. The pond water seemed cool to swim in when the weather was warm, and warm when the weather was cool. The pond was a *stable thermal storage mass,* a *heat bank,* which took days to adjust to a change in ambient air temperature. Diving to the bottom, we found the coldest temperatures. The warm water was at the surface, absorbing heat from the sun. The water was stacked in *thermal layers.* On hot summer days, when the silo was empty and the door at the bottom was opened, air rushed in and up through the silo in a *chimney effect, thermosiphoning* cold air up and out the top as the warm air above pulled the cold air up from below.

I started my first year in architecture school at Yale in 1962. It was the best of times. The place was buzzing with ideas and energy. The foremost practitioners of contemporary architecture were brought in on a regular basis to challenge and encourage our talents. It was a time of great idealism and ambition. No limits were implied. We would graduate and design great buildings—simple as that.

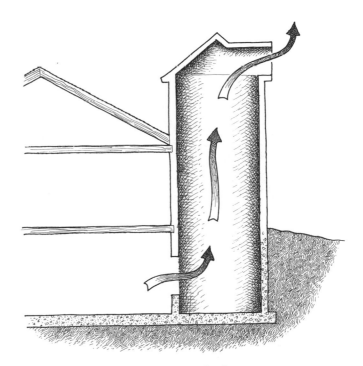

Thermosyphoning on a hot day on the farm.

New tools of the trade for architects.

Learning to Build

Although a high premium was placed on pure design at Yale, another trend was begun during those years—unprecedented and long overdue—the trend toward learning how to build what you'd designed. Most architectural education has not included nuts-and-bolts building experience in its curriculum, and lack of hands-on experience in the field has been an unfortunate liability for too many architects. Reading about 2 x 4s is no substitute for having nailed a few thousand into place.

The impulse to learn about the real world of building was acted upon and dramatized by a charismatic second-year student, David Sellers. He took a year off to build a house he had designed for his brother. His periodic visits back to the ivory tower left many of us envious of his opportunity. It seemed that whatever we might learn in school would be incomplete until we could know what it was like to personally put all the parts of a building together. Combining the conceptual and the practical, combining art and craft, was certain to improve our understanding of how to design a building.

And, of course, it did. One by one we followed his lead, summer projects, part-time jobs, leaves of absence—anything to swing a hammer. Within a few years, the faculty was convinced we were on to a good thing. One of the first-year classes was asked to design a community building for an Appalachian town. When the design was completed, the class left

The Fallingwater house, by Frank Lloyd Wright, makes use of a brise soleil

for Kentucky and built the building from the ground up. The design-build concept was officially pronounced a success. But, more importantly, essential groundwork had been set. Those attitudes and the practical approach to problem-solving became essential to creative thinking when building design was confronted with the implications of the oil embargo in 1973.

Conservation Ignored

In the Modernist, pre-OPEC architecture, issues of energy conservation were either ignored or gratuitously manipulated as devices for architectural expression.

Le Corbusier's *brise-soleil* (an extension of floors and walls beyond the glass line) may have worked well as a shading device, but it was most importantly heroic sculptural form, validated by the aesthetic rather than the practical. By the early 1970s Modernism had exhausted the potential of its credo "form follows function" and was discredited for its lack of historical, regional, and cultural sensibilities. In a flurry of instant history-making, the vacuum created by the demise of the Modernist movement was filled by the self-appointed opportunists of the *Post-Modernist* movement. Post-Modernism proposes an architecture of self-indulgent, eclectic confection, cipherable only to its privileged practitioners—who insist the Emperor *is* wearing

11

new clothes. Issues of energy efficiency are ignored. The Post-Modernist issues instead tend to be literary and a-temporal, orbited in a semantic stratosphere of its own making. They are cerebral, philosophic, and engaging, but lopsided in favor of artfulness. The glossy magazines love the debates and wait to see in which direction the hemlines will move next year.

Architecture is unique among the arts in that its purpose is both spiritual and functional. At its best, its form is a mirror, perceptive and harsh, reflecting the ideas and truths of its era. It is, therefore, more than just aesthetic expression or practical shelter, but a combination of the two. The Post-Modernist movement may be important insofar as it seriously addresses some of the issues, but its potential won't be fully realized until it embraces the hard global issues of our energy-scarce times.

New Architecture

Meanwhile, in the not-so-glossy magazines, in the not-so-Post-Modernist offices, a new architecture is emerging. It began with grassroots tinkerers and engineers, clumsily shoehorning the new solar technologies into drab boxes. Flat plate collectors were pasted like postage stamps on ranch house roofs, and earth-sheltered houses were stuck into hillsides like forlorn bunkers. As the technological shakedown continued, the performance data were encouraging. This new stuff actually worked as they said it would! But it was

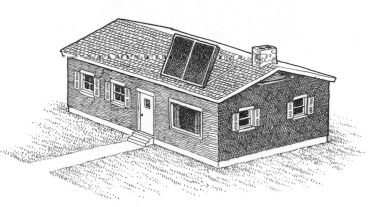

Solar comes to the ranch house.

unprepossessing shelter and hardly qualified as architecture. The machinery was working, but it hadn't found its expression in form and space. But, as the seventies drew to a close, a decade of promise dawned with dozens of exciting buildings that combined the art of architecture with the craft of low-cal design.

In the chapters to follow, we'll examine the workings of earth-sheltered buildings, active and passive solar heating systems, double-envelope shells and superinsulated structures. Their strengths and weaknesses will be analyzed, and proposals for a synthesis of their many virtues will be made. The discussion will be low-tech and practical, with an emphasis on how and why. The single-family residence will be used as the working model, and references to the agony and ecstasy of the creative impulse will be kept to a minimum.

HORSEPOWER

By 1850 almost half of New Hampshire had been cleared for agriculture and grazing. Sheep were everywhere and it wasn't until the Australian-grown merino wool undercut the New England market at the turn of the century that the forests crept back in. Today only 15 percent of New Hampshire remains open. Without grazing, the hard-won pastureland succumbed to the inevitable cycle of poplar, cherry, grey birch, and white pine, evolving decades later into the classic northern climax forest of yellow birch, sugar maple, and American beech. Hemlock crowded the damp north slopes, and black spruce peppered much of the thin, overgrazed topsoil that had probably never been worth clearing in the first place.

Clearing land in 1850 was hard work. To begin with, much of the country was virgin forest. The wide pine floorboards and paneling so cherished by today's antique buffs were simply the practical result of sawing up huge trees. These giants were cut down with double-bit axes and two-man crosscut saws. What might take a chain saw operator fifteen minutes to accomplish today took two men an hour to complete back them. Teams of oxen twitched the logs to a sawmill at a ponderous pace. An accumulated ten miles a day was exceptional. Modern four-wheel-drive skidders move at ten miles an hour. And most of the trees cut never reached the mill. They were burned on the spot, the vast stores of potential energy held in their fibers converted to

PHOTO FROM THE BETTMANN ARCHIVE, INC.

In 1850, they did it without chain saws.

*Measuring heat loss and heat gain in a
lived-in house is a tricky proposition.*

heat by fire. Winter landscapes were smudged
with smoke as the settlers cleared and burned
their way into the frontier.

An Energy Vocabulary

When we talk about energy, we use a vocabulary that describes a capacity for doing work. In its simpler, archaic form, the horsepower was an appealingly graphic term, a unit of energy understood by virtually everyone in preindustrial America. A horse could carry a rider forty miles a day and travel thirty miles an hour in short bursts. A team of husky Percherons could pull up to 12,000 pounds across flat ground, and a team of clever workhorses pulling a hay rake could gee and haw its way around a sidehill hayfield all day with uncanny agility.

Today a horsepower represents a measure of mechanical work, a stipulated unit of energy. In preindustrial terms, the standardized unit was irrelevant. Horses were individuals with individual capacities. Cross-reference was unnecessary. Because there was no mass production of work-producing machinery, work-producing animals—and men and women—were evaluated individually according to the promise in their firmness of muscle and length of tooth.

The steam engine and electricity changed the way in which work was defined. A more refined language was needed to describe work in its several forms: mechanical, thermal, electrical, and chemical. Not only did each category develop units of standardized measurement, but these units were transformable within the overall definition of energy. One horsepower-hour equals 2,545 Btu. One watt-hour equals 3.41 Btu. One Btu equals 252 calories. There are fifty calories in a head of lettuce. Translated into calories, the energy required to produce one head of lettuce equals about 400 calories. Shipping the lettuce from California to Boston consumes about 1,800

How much energy was used to get that lettuce to your house?

calories. In terms of energy efficiency, lettuce is a loser. But with bacon and tomato it makes a nice sandwich.

The Btu

In the discussion of energy (heat)-efficient building design, the Btu is the most common point of reference. It is typically expressed within the context of a unit of time—Btu/min, Btu/hr, Btu/year—in or within the context of a unit of volume in Btu/gal, Btu/lb.

Measuring Loss and Gain

Measuring heat loss and heat gain in a lived-in house is a tricky proposition. As performance becomes increasingly critical, the criteria seem to multiply and the increments of measurement are decimaled out another point or two. Engineers welcome the opportunity to add and subtract every loose Btu they can find. But there are many areas that can't be precisely quantified and are often ignored. The slide rule can't account for thousands of Btu lost or gained according to peculiarities of building usership. For instance: how much waste heat does the refrigerator produce? How many times a day is the front door opened and for how long? Is a wood-burning stove used? How long, and what kind of wood was burned, and how wet/dry was it? (Wet wood is up to 25 percent less heat-productive than dry wood.) Was the exhaust fan left on/off in the bathroom/kitchen? For how long? How accurate was the thermostat and was it really 65 ° F. all over the house or was it 55 ° in the bedroom and 70 ° in the kitchen? Do the storm windows fit? Is the six inches of insulation in the wall wet or compacted? Is it one inch short here and there? Does the spruce tree on the front lawn shade the south window in January? How much sunlight actually gets into the house? (Sunshine availability can vary 25–30 percent above or below "normal" from year to year.) Are dirty windows or screens reflecting away 15–30 percent of the incoming sunlight? Is the backfill around the foundation wet or dry? (Wet backfill conducts heat away from a building.) How about the wind on the site? And, finally, how is performance eventually measured?

Owner testimonials are sincere but subjective. A recent article in an energy-oriented monthly quoted the owners of a miraculous new house . . . "We used no fuel at all for space heating. . . ." Two sentences later, they mention, in passing, that they used three cords of wood to fire their cookstove.

No one wants to admit that the extra $15,-000 spent on a whiz-bang Solar-Saver-System would be more productively invested in municipal bonds. But, in fact, that's often the case.

The Answers

Most of these questions are rhetorical in nature and their answers are found in wisdom rather than science. I mention them partly to

Even trees will change the Btu demands of your house.

needle the computer jocks and partly to ask the qualitative question. The perfectly engineered, operated, and monitored house may perform like a Swiss watch as long as its inhabitants are equally precise in their movements. But life is more than a checklist. What is the house like to live in? How many Btu do you trade for a view? How do you compare the price of a gallon of oil to the satisfaction of a perfectly composed space? Is life with an oil-drum living room wall a step forward—or to the side?

Horsetrading

Horsetraders deal with a complex commodity. Their ability to quantify what they see— or they think they see—is the critical measure of their skill. Sire and dam are important, but movement holds the code. "Trot 'em around again, Bill, break 'em into a canter for me." He's over-in-the-knees but he's a broad-chested little bay. Does he "wing?" Check for splints, for cow-hocks, sickle hocks, for sound pasterns, and thrush. Is he cold-backed? Does he chip in and hang his legs over a jump? What about those lop ears? Does he crib or weave in the stall? Inspect the eyes for moon blindness and the capped elbows for infection. How's he look? Appaloosa, strawberry roan, Palomino, chestnut, old gray mare. Star on the forehead? A pretty blaze, three white stockings and a full mane and tail. He'll never trade if he's goose-rumped or ewe-necked or pig-eyed or bog-spavined or parrot-mouthed or just plain workaday ugly. A horse is a living machine and a thing of beauty, an individual like and unlike all other horses. As an instrument of work he's predictable but subject to changes in health, age, and care. As a living creature he exalts an elemental sense of beauty, grace, and courage central to the history of man.

What's a Horsepower?

In terms of mechanical work, one horsepower equals 33,000 ft/lbs/min—the capacity to lift 33,000 lbs. one foot in distance over a period of one minute. When applied to a piece of machinery, the standard is easily enough verified. But the power of an organism is multifaceted, and to judge it by one standard alone is a mistake a good horsetrader would never make.

The Challenge of Design

I look at a house the same way that I look at a horse. If I had the opportunity to design a *horse,* the conformation would be the easy part. The animal's spirit, its temperament, its personality would challenge me most. And so it is with designing a house. The *quantitative* aspects are just that, legible and predictable engineering. The *qualitative* aspects are far more mysterious and rewarding in ways which, of course, defy measurement. The ideal design solution has full exposure to both. The qualitative side we'll call architecture and leave to a later discussion. The quantitative side we'll analyze now as part of the background for evaluating energy-efficient houses.

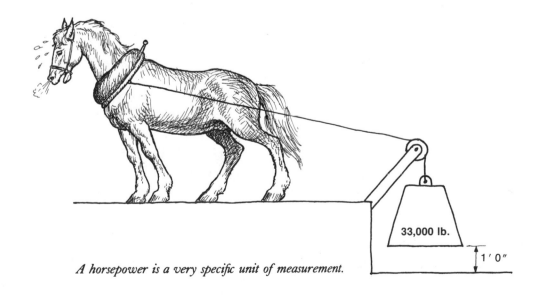

A horsepower is a very specific unit of measurement.

33,000 lb.

1' 0"

degree-day figures for various parts of the country:

Burlington, Vermont	7,870
Washington, D.C.	5,020
Chicago, Illinois	6,130
International Falls, Minnesota	10,550
New Orleans, Louisiana	1,470
Phoenix, Arizona	1,550
Denver, Colorado	6,020
San Francisco, California	3,040
Seattle, Washington	5,190
Fairbanks, Alaska	14,340

Efficiency of Buildings

The two basic criteria used in evaluating a building's operating efficiency are climatic environment and shell insulation. In other words, how hot/cold it is outside, and how well do the walls/roof resist intrusion of the outside temperatures? Climate is evaluated in three basic areas: temperature, solar radiation, and wind chill factor. Temperature is expressed in degrees Fahrenheit or Celsius.

Temperature is also the key to the heating degree-day index. The heating degree-day index is a useful rule of thumb for comparing one site to another and for calculating energy consumption. It is based on the assumption that a 65° F. outside temperature will impose no heating requirements on a building. Each degree below 65° F. accounts for one degree-day, with the average temperature figured by adding the maximum and the minimum temperatures for the day and dividing by two. If, for instance, the average outside temperature is 35° F. each day for one week, or 30° below 65°, calculate 30° × 7 = 210 degree-days for that week. Added up for the entire heating season, the number of degree-days indicates the total heating requirements for that period. (Cooling degree-days tell the same story, measuring the amount of air-conditioning requirements a hot climate would impose upon a building.) Some typical heating

Measuring Radiation

Solar radiation is measured in langleys, joules, or Btu/ft^2. (One langley per minute equals 221.2 Btu per hour per square foot.) In the outer atmosphere, solar radiation is measured at a constant 432 Btu per square foot per hour. By the time solar radiation has reached the earth, its potential has been reduced to about 150 Btu/ft^2/hr. of usable heat. This figure may be reduced further due to rain clouds, dust, and smog. Solar radiation does not alter degree-day calculations, but it can contribute substantially to heating/cooling requirements. Phoenix receives an average of 1,740 Btu/ft^2 per day in the winter months. Burlington averages 680, Denver 1,640, and rainy Seattle only 490. The heat in one square foot of sunlight per winter day in Seattle is equivalent to the heat given off by a

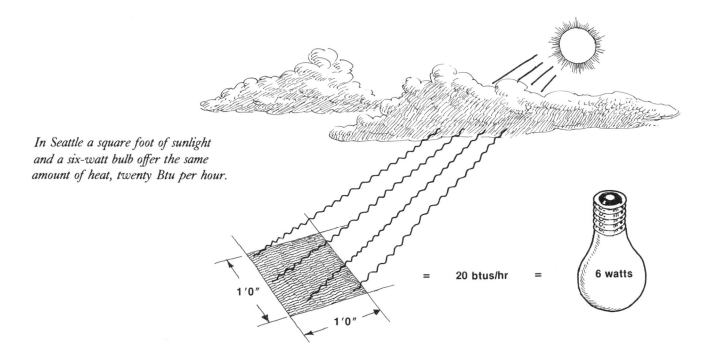

In Seattle a square foot of sunlight and a six-watt bulb offer the same amount of heat, twenty Btu per hour.

= 20 btus/hr = 6 watts

single 150-watt light bulb burning for *one hour*. Over twenty-four hours, Seattle receives about 20 Btu per square foot—equivalent to the heat generated by a six-watt bulb over the same period of time.

Phoenix receives 3½ times as much solar radiation as Seattle and has less than a third the number of degree-days. The combination of low heating degree-days and a high solar radiation index makes Phoenix an ideal location for low-demand, solar-assisted heating. Seattle's access to hydropower offsets its solar liability and stresses the wisdom of a flexible, regionally responsive national energy policy. Decentralized power grids allow maximum utility of regional energy potential—fossil fuels, wind, solar radiation, hydro, geothermal, or biomass—without incurring the substantial transmission inefficiencies inherent to the mammoth centralized systems.

Free Energy Source

The argument for solar radiation heating is appealing because the energy source is free. Even in Seattle, a 500-square-foot active col-

lector might bring in an average of 245,000 Btus per winter day, accounting for between a third and a half of a well-designed home's heating load. Well, the energy may be free, but the expense of harnessing it is not. An active hot air system—collectors, ductwork, hardware, and storage—costs between $20 and $25 per square foot of collector in 1981. Ten to twelve thousand dollars! At 10 percent interest, the $12,000 spent on an active system could earn $1,200 a year, three to four times the cost of the fuel saved.

Tax subsidies favoring the active systems enhance the proposition slightly, but not enough to confirm cost effectiveness. And the scenario of $5-a-gallon fuel oil suggests only that the energy-intensive collector components will become equally costly, putting the systems' costs in the $40,000 bracket.

This brief digression into the economics of active solar radiation systems is only meant to suggest that:

1. Seattle is not an ideal location for cost-effective use of active solar systems, and,

2. Traditional flat plate active systems are at a disadvantage because their high capital

A live-in collector serves many purposes.

outlay costs are not justified by practical performance alone.

If, on the other hand, the 500 square feet of glazing on our Seattle example was the south wall of an accessible, plant- and people-filled room, the cost analysis takes on another dimension. The "collector" is usable. It is valued beyond the purely practical. The extra money spent on gathering sunlight gratifies the eye and the sense of architectural experience as well as saving fuel. The live-in "collector"—sunspace, greenhouse, atrium, sunporch—is a good example of an *integrated,* engineered, and architectural solution to solar-sensitive design.

Wind Speed

Heating (or cooling) degree-days and the solar radiation index are the basic criteria used to express climatic character. In some locations, wind speed must also be reckoned with. The back-to-the-land impulse begun in the counter-culture sixties included among its ideals a strong mandate for self-sufficiency. "We'll have our own garden, a cow, wood heat, and a *windmill.*" In most places the granola got grown, the cow got milked, and creosote choked up the chimneys, but the windmill was a bust. The windy areas of the United States are not particularly hospitable to nouveau agrarians. The coastal regions are notably windy, but overpopulated (and therefore expensive), and the harsh, ever-windy Great Plains run counter to the romantic notion of woods-water-peaceful valley seclusion.

Windmills don't begin to operate effectively at less than ten miles an hour, and a twenty-four hour, ten-mile-an-hour average is unusual in most of our habitable areas.

Wind Chill Factor

Windmills aside, the wind *chill* factor is a significant design consideration in assessing climatic effect on buildings. When there is no wind around a building, a film of air hugs the outside surfaces, acting as a thin, hovering layer of insulation. As the wind speed increases, this layer is drawn away and heat is drawn through the wall at an exponential rate. The chilling effect of a windswept building site can be countered with fences, fin-walls, window planting, extra insulation, and earth-sheltered design.

A fifteen-mile-an-hour wind on a 20° F. day creates a wind chill factor equivalent to −5° F. still air temperature. The same wind in 0° F. air has the effect of −31° F. still air temperature. A site with an average wind speed much over five miles an hour should have the wind chill factor impact included in heat loss calculations. The effect of the impact is to lower the ambient outside design temperature according to the amount indicated by wind speed and to anticipate an increase in infiltration and air changes due to wind pressure.

Wind speed is easily measured with a recording wind anonometer. The anonometer is a small vertical-shaft device with wind catching cups or blades which spin the shaft.

Wind chill, simplified.

The shaft's revolutions per minute are translated into miles per hour and recorded digitally. The device should be run for as many months as possible in order to indicate a true long-term average. Anonometers are available on a rental basis from windmill manufacturers or for sale through companies advertising in solar publications.

Temperature, sunlight, and wind are the critical elemental forces defining the climatic environment around a building. Once they've been evaluated, the building can be designed to respond to its environment as advantageously as possible. Keep the cold out in winter, the heat out in summer. Draw in the sunlight and protect against the wind.

Insulation

Windows and shading devices regulate incoming sunlight. They also form a substantial part of any architectural vocabulary. Insulation, on the other hand, is purely functional and makes little demand on aesthetic expression. Insulation is the foremost component of energy-efficient designs. The first $1,000 added to any building budget for extra energy conservation purposes should be spent on insulation. And probably the second and third thousand as well. Before any of the new solar technologies, the old principle of heavy insulation pays off first.

Reindeer hair is hollow, providing extra trapped air spaces for added warmth. Down feathers keep a bird warm because the miniscule cavities between the down are filled with trapped air. (Oil spills kill waterfowl, not because of toxicity but because when feathers become soggy with oil, a bird has trouble floating and has no protective insulation against the cold.) Building insulation operates on the same principles; it must contain air and the air spaces must be kept dry. Water is a good conductor but a poor insulator. Soggy insulation simply doesn't insulate.

Three Kinds

Building insulation is fabricated in three principal forms:

1. *Fiberglass batts* are typically dimensioned to fit between framing members sixteen or twenty-four inches on center. The batts range in thickness from 3½ to 12 inches. The batts are covered on one face with kraft-type paper or reflective foil, or have no covering on either face. R-values (an index measuring thermal resistances, with the higher numbers indicating greater insulation) run from R-12 for 3½-inch batts to R-38 for the 12-inch batts.

2. *Rigid foam board* insulation is made in two generic categories: urethane and polystyrene. Both are called "closed-cell" foams and are designed to resist absorption of water.

The urethane foam is rated at around R-8 per inch of thickness. It's available in 4 x 8-foot panels from ½ to 2 inches thick. It's manufactured with foil or paper bonded to both faces and is the most expensive, per R-value, of the insulations listed. If burned, it emits a toxic and possibly lethal gas. As it ages, its R-value decreases by as much as 20 percent.

Polystyrene foam board is the most widely used rigid insulation product. There are two types, *extruded* and *expanded.* The extruded polystyrene has an R-value of 5.4 per inch of thickness and is the most proven and truly closed-cell variety of foam board insulation. It's fabricated in 2 x 8-foot panels one to two inches thick with square or tongue-and-grooved edges. Expanded polystyrene has an R-value of 3.4 per inch and is the least expensive per R of the foams listed. It is made in 2 x 8-foot panels from ½ to 6 inches thick.

3. *Loose fill* insulation is designed to be poured or blown into place. It requires the least amount of labor and is generally the least expensive of the common insulation products. Cellulose fiber is the most practical and economical with an R-factor of four per inch. Superinsulated houses are typically insulated with cellulose fiber because it pours so easily into the thick wall cavities and costs a little over one cent per R per inch. Other forms of loose insulation include expanded mica (vermiculite) and chopped fiberglass. Both are acceptable, but more expensive, substitutes for cellulose fiber.

Animals adapt to the changing seasons by growing or shedding hair, hibernating, migrating, and adding insulating layers of fat to their bodies. The fur-bearing species depend on hair for cold-weather survival. But the fur doesn't grow infinitely long and thick at the

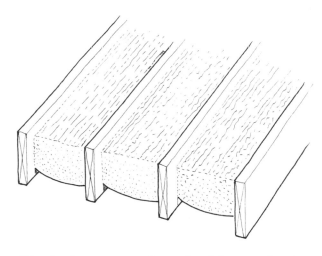

Fiberglass batts are manufactured to fit between framing members.

outset of cold weather. A genetic mechanism controls the amount, type, and length of hair grown so that the animal's overall comfort is best served. Winter's temperatures are widely variable. Too much hair (insulation) could overheat the animal. Too little would be hazardous. The animal's requirements are complex and kinetic and provide a useful analogy to the insulation requirements of a building.

Create Own Heat

Mammals create their own body heat by the metabolic conversion of food to heat. As the body's furnace works harder to keep the body warm, more food is consumed. The primary factor determining thickness of fat (fuel) and fur (insulation) is the availability of food (energy). When food is scarce during the winter, animals rely upon heavy insulation rather than food for increased body heat. The natural economy of survival dictates the relationship between energy consumed and energy conserved. Simplistic as it is, the principle applies to our sophisticated post-industrial society with the same degree of urgency as it does to the caribou north of Yellowknife.

Insulating a building costs money. The expense is found not only in the cost of the insulating materials but in the cost of the added shell thickness required as well. For example, 3½ inches of fiberglass fits perfectly into a typical 2 x 4 stud wall. Adding more fiberglass, however, means a wider wall stud. Structurally, the 2 x 4 may be sufficient, but a 2 x 6 or 2 x 8 stud system provides the increased space required despite being structurally oversized (and therefore uneconomical). Aside from insulation and added costs of extending standard window and door jambs (to accommodate the extra wall thickness), the reduction of usable floor space by four inches times the building perimeter must be included in a cost-effective evaluation. A two-foot thick wall stuffed with insulation will never pay for itself in fuel savings.

On a strictly cost-accounting basis, we can make some sound judgments on how much insulation should be used in various parts of a building. The difficult part of the equation is predicting future fuel costs. How much will heating oil and other fuels cost five or ten years from now? If we are overly pessimistic, our insulation expenses will be economically unjustifiable. The temptation, of course, is to say "the more insulation the better," but if we mean to design within the real constraints of the real world, there comes a point where enough is enough. The chart lists insulation types and cost per "R" of insulating value. In general, this chart is useful in primary planning, but should not be interpreted literally as the coming chapters will suggest.

COST EFFECTIVENESS
OF VARIOUS INSULATIONS

TYPE OF INSULATION	R-VALUE/INCH	1981 COST/BD FT	COST/R/INCH
Cellulose fiber (loose)	4	4.4¢	1.1¢
Fiberglass batts, foil faced	3.2	5.7¢	1.8¢
Urethane (isocyanurate)	8–20%±	55¢	6.9¢–8.6¢
Extruded polystyrene	5.4	37¢	6.8¢
Expanded polystyrene	4	15.5¢	3.9¢

3

EARTH-SHELTERED

When the need for energy-efficient buildings became urgent in the early seventies, the design community looked over the prospects, chose up sides, and went to work. The high-tech approach made a dazzling but brief appearance, soon stunned by its own economic extravagance, tripped by its own fancy technological footwork. The early, exotic systems tickled our imaginations but left us disappointed with cost overruns, leaks, warping, and Rube Goldberg design.

But the race was on and the press was keeping score. Editors demanded "energy-interest" stories. "We Heated Our Home for $1.32" was the headline everyone was looking for. Solar voodoo was practiced openly, adding enough high-tech patois to cocktail party patter to make an eco-elitist shiver clear down to his Vibram-soled hiking boots. Factions emerged. New religions were formed; alliances were broken and regrouped around the latest solar shuffle. Each solution out-conserved the other and the misuse of energy took on a moral dimension that would have pleased even Calvin. Eutectic salts, sun-tracking parabolas, and photovoltaics headed the list of technological triumphs whose time will come, but the dawning of the Solar Age needed a more prosaic promise to insure its future.

Within the last decade, building costs have almost tripled. Operating costs (heating and cooling) have increased five-fold along with similar cost increases in maintenance and repair. Anyone who can afford to build today is bound to be looking at the long-term efficiency of the structure, and is invariably willing to invest considerable amounts to protect against exponentially rising energy costs.

Three Approaches

In single-family house design, the eighties began with three substantial schools of thought, backed by almost a decade of trial-and-error practice and prototype experiment. The earth-sheltered, double-enveloped, and superinsulated concepts emerge as the three most credible hypotheses of high performance design. Each has considerable strengths; practical applicability, economic merit, proven performance, and aesthetic potential. They also have their faults and faulty premises. An explanation and analysis of the three seminal types will show *how* they work and *why*. With these issues clarified, the design of a *Superhouse* can hope for a three in one solution that will combine the best of the best and be done without the tagalong inconsistencies that forever plague the monocausal solution.

The Deep, Dark Unknown: Escaping the Underground Image

The idea of living *in* the earth gives us pause. With few exceptions, man has chosen to live *on* it, move across it, not through it. We dig it up, plant it, harvest, shape, and defile it. We cherish it as a symbol of *patria* and inheritance. We sell it, buy it, and build on it, but when we contemplate living *in* it, our traditional expectations of house and home are challenged with the uncomfortable imagery of the deep, dark unknown.

Our literature provides substance to our subterranean angst. Tom Sawyer, Alice in Wonderland, Peter Rabbit, and Tolkien's Hobbit all celebrate the mysterious other-worldliness of life below. The correlation between being in the ground (interred) and death is an indelible cultural reflex. Death is, after all, most typically followed by burial, the grave, the vault, the crypt, the tomb, all underground, hidden, mysterious, and final.

More practically, our everyday experience tells us that basements are dark and dank. Tunnels and caves are known to ooze moisture. Bats fly between the stalagtites. Rats scurry between the stalagmites and we imagine the hugger-mugger man lurking in every corner. Our childhood cellars were invariably full of cobwebs and titillating fear. The French call the basement *le cave*. Some Americans turned theirs into bomb shelters in the fifties. Most were left to the storage of coal, potatoes, and out-of-season bicycles. Mystery story writers use the basement floor as a discreet repository for victims of foul play. Folk tradition has the basement stored with homemade cider and wine, consumed by lantern-light in paralyzing quantities by perspiring, thick-waisted stepfathers with vile tempers and dark, brooding minds.

Recent efforts to salvage underground space from the tyranny of darkness sponsored

Early earth-sheltered housing.

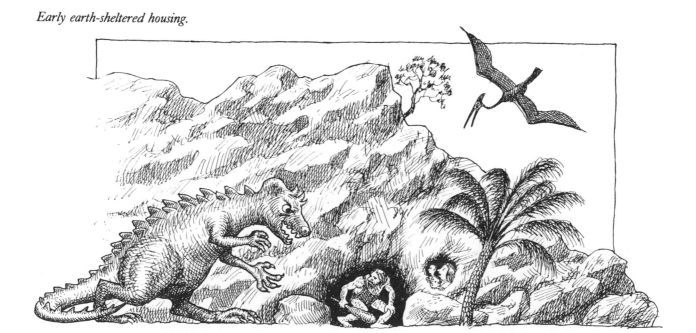

the phrase "daylight basement." This feature carried the assurance of a separate entrance "down there." If you consented to go *down,* you could also go *out.* Hence, the phrase "down and out" took on new meaning.

Our basements *were* dark and uninviting for several reasons. Houses were meant to be seen as a symbol of status and achievement. House pride. "A man's home is his castle," "the big house on the hill," "the ivy-covered cottage," conspicuous emblem of success. The inconspicuous part was the basement, designed and built to be just that and no more. The old basements were laid up with fieldstone, with no waterproofing, no footing drains, and usually a hard-packed dirt floor. Moisture and cold were acceptable as long as the basement stayed "down there."

The basement, traditionally the "bad" area of a home.

Radical Proposal

The underground house movement had to overcome a host of anxieties, both real and unfounded. Living underground is a radical proposal, a more extreme departure from the American ranch than superinsulation, greenhouse-sunspaces, or double-envelope construction. But the earth-sheltered houses proved themselves early on, and the pioneers spread the word: the houses were dry, light, airy, and energy-efficient, warm in winter, cool in summer.

The design of energy-efficient earth-sheltered buildings is based on a simple premise, a simple understanding of the nature of *earth.* Because of its relatively moderate and stable thermal properties, earth is an economical and permanent means of sheltering a building against extreme weather conditions.

The premise is sound, and modern technology has provided the requisite materials and techniques. Structural design analysis for these buildings has always been straightforward, and waterproofing, the main area of anxiety, has been thoroughly demystified and perfected. Quantitatively and qualitatively, the traditional problem areas of earth-sheltered buildings are now identifiable, predictable, and controllable.

Limited Use

I don't suggest that all architecture be earth-sheltered. In fact, the majority of it *shouldn't* be. Large-scale urban projects, for instance, requiring natural light and sight (high density housing, office space, and hospitals) are inappropriate targets for the earth-sheltered concept. High water tables, poor soil conditions, and certain preceding architectural environments also make the concept inappropriate.

On the other hand, theaters, stadiums, convention halls, warehouses, retail outlets, and light manufacturing plants would suffer no penalties in trade for the substantial benefits of earth-sheltered design.

At the present, our universities and rural, single-family dwellings account for most of the interest in the earth-sheltered approach, but other uses are bound to follow. The engineering aspects are proven, the sociological

impact is in the making, and the architectural significance awaits definition by the architects themselves.

Aesthetics

The question of architecture in these buildings is the most elusive and challenging of all, and yet it's been addressed the least. With few exceptions, the emphasis has been on the technological elements only, almost as if good engineering precluded good design. Every well-engineered but tasteless earth-sheltered structure built simply fulfills the cynics' prophecies that only woodchucks, not people, are meant to live below ground.

As in the study of any building type, the vital distinction between a unit of building and a work of art is essential. A warm, dry habitable space is not necessarily a work of architecture. (Neither, regrettably, are some works of architecture necessarily warm, dry, or habitable.) But the two are not mutually exclusive. The earth-sheltered approach is as fundamental as any in architecture, as old as the art itself. It's as valid in a vehicle for the expression of a new architectural vocabulary as any contemporary issue—in fact *more* valid than many.

Architecture is the most omnivorous, pervasive, and enduring of the arts. We are consumed by architecture. Not only is it around every corner, it defines the corner itself. We go to look at historic paintings at the National Gallery and the building steals the show. Ar-

An earth-bermed passive solar house in Santa Fe, New Mexico, designed by David Wright.

chitecture is a barometer of cultural change and a receptacle of all forms of human enterprise. It is at once the most utilitarian and spiritually ennobling of all the arts. From the sophisticated technologies of the high-rise tower to the awe-inspiring ambiance of the Gothic cathedral, architecture provides the vital context within which our important ideas and events have grown and flourished. Architecture combines engineering and aesthetics in a way that is unique among the arts.

Engineering aside, the architect's creative powers are, as with any artist, mysterious, personal, and elusive. The origin of artistic ability is difficult to identify and even more difficult to duplicate. We can teach physics to just about anyone, but only a few can write good poetry, and even they don't really know how they do it.

The issue of aesthetics is a difficult subject, full of rhetoric and definitions of definitions that all spiral back to the only question worth asking: "Do you like it or don't you?" Unfortunately, many earth-sheltered buildings haven't provided enough visual substance to promote the question. The underground movement has come a long way in the last decade, but the least exploited of its many virtues is its potential for aesthetic expression.

Symbolic Meanings

There are surprisingly few cognitive forms in architecture. These forms have evolved and survived because they make sense to our hearts and minds.

And because they have survived and become identified with particular functions, they have become infused with symbolic meaning. The column, the colonnade, the lintel, the

PHOTO BY MARILYN MAKEPEACE

The entrance must invite and excite.

arch, the dome, the gable, the step, and the tower—these forms signify certain expectations which we learn to anticipate. The medieval fortress, the Renaissance stairway, the Moorish atrium, and the Madison Avenue skyscraper have come to symbolize strength, ceremony, sensuousness, and industry. A mere building provides shelter and security, but in order for that building to become architecture it must "signify" something in a familiar, visual language.

Earth-sheltered buildings represent an infinite variety of programmatic functions—commercial, industrial, institutional, and residential—and each one must signify its purely functional characteristics in whatever way possible within the earth-sheltered context. What we're looking for is an architecture vocabulary of "earth-shelteredness" that signifies the unique dynamics of building underground.

Much has been made of the rounded, maternal, female forms of the subterranean Minoan structures. The evocation of "Mother Earth" is too tempting to resist, and the womb-like forms of these "primitive" buildings capture the essence of what we seek underground: warmth, security, and strength. If these ingredients do indeed characterize the impulse to "dig in" (and I believe they do), then how does the architect express them? How do we signify warmth or strength, for instance, with masonry, wood, and glass? The answers are familiar and intuitive, part calculated and part inspiration.

Earth-sheltered buildings are safely engineered to support tremendous loads of earth.

We may know this rationally, but subliminally, we need to see how it's done; we want to be reassured that the building is safe and sound. We need evidence of a muscular structural system as a means of signifying this dynamic challenge to gravity. The underside of a flat concrete slab suggests nothing about its thickness or how it was made, and gives no impression of the function it performs. On the other hand, massive timbers, ribbed or vaulted arches, open trusses, or articulated concrete planks establish credibility and a convincing sense of security. The exposed structural components not only demonstrate their heroic function, but we see them as "beautiful" because they are arranged in a way that makes sense and comforts us.

The Curved Wall

While the arch is the strongest shape for resisting vertical loads, a curved wall (an arch on its side) is equally superior for resisting lateral loads. (Again we return to the Minoan womb, where curved walls both structurally and symbolically signify strength and security.) A further sense of mass and solidity is perceived through window and door openings where the thickness of the walls indicates their strength and security.

The entrance to an earth-sheltered building might convey the nature of what's inside, and how the inside might be expected to relate to the outside. The point of entry must extend an invitation to a special environment, with hints of what's to follow. If we can see the earth on the roof, or earth leading up to the roof, we're primed for the ensuing experience. If we're provided with a glimpse of the structural system, we're prepared for the excitement of seeing how it's done, how all that earth is held in place. And if, from the entry, we can see inside to another source of daylight, once again we're intrigued and drawn into the building to find out how it works. The point of entry must invite and excite, a preview of things to follow, not just a door in a wall like any other door in any other wall.

Materials

Materials figure importantly in our perception of our environment. Stone and brick connote a solidity that Formica and wallpaper don't. Carpet disguises, quarry tile convinces, gypsum wallboard is anonymous, plaster suggests integrity, and wood conveys warmth. A material that shows us how it's put together and what it's made of enriches our participation in our environment. In the underground environment, where structure is so fundamental to every design decision, a celebration of building material is a critical aesthetic priority.

Form, space, and light are basic to all architecture. Underground, these elements must be manipulated in such a way as to recognize and capitalize upon the "undergroundness" of the structure. The forms of underground structures will first reflect the loading from above, and secondly, anticipate the structural openings towards outside sight light and sight.

"A convincing sense of security."

PHOTO BY DAVID MARTINDALE

Space within the forms invites variation—narrow, wide, low, high, contiguous, and compartmentalized. A row of identical boxes inside a structure misses the opportunity for enrichment. Differentiation of spaces performing different functions is essential.

Angle of Lighting

Natural light is ever-changing and important to our psychological sense of well-being. Generous quantities in the underground building are a must. Horizontally introduced light is preferable to light from above. The horizontal line of sight is our natural aspect whereas light from above adds to our sense of being "down under." Light keeps us in contact with changes in the natural world outside and reinforces our sense of womblike security in a way that artificial light can never do. Lighting from natural sources should be thought of as an important design tool, both practical and dramatic, essential to the exploitation of space and form.

In brief, the underground aesthetic defines a way to symbolize the warmth, solidity, and security of underground living. The expression of structure is central to the thesis, while space, form, and material will complement the structural principle. The issue is intriguing and the solutions are developing. With courage and talent, we'll soon see beauty—below ground—where we've never seen it before.

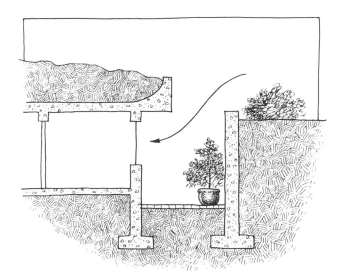

Horizontal light is preferred.

Vermont above ground, Georgia below.

Impact

It's difficult to argue that even the simplest hut has ever benefitted the natural environment. While the railroad, the pipeline, and the parking lot appear to signal our progress as a species, their environmental impact is always measured in terms of how much *harm* they do. We will, by necessity, continue to build, but our priorities must change if we ever expect to approach a symbiosis with nature.

In the last decade, the combined exponential rates of population growth and resource depletion have chillingly emphasized the finite nature of our enterprise here on planet Earth.

Earth-sheltered buildings have been around for centuries. Their recent popularity is due to their appealingly simple concept, positive environmental impact, low-tech construction methods, and energy-wise performance. They're here, and they work.

Surrounding three sides of a house with a Vermont hillside is, in effect, surrounding that house with a climate found as far south as Georgia. The earth's temperature, eight feet down, can be more than 70 ° F. warmer than the coldest 30 ° F. below zero Vermont February night. Add a wind chill factor to the 30 ° below temperature and the case for earth-sheltering is complete.

Ideal Site

The ideal site for a single-level earth-sheltered home is a one-to-five or one-to-six slope oriented to within 15 ° of due south. Access to it from the north, east, or west is preferable, allowing the southern side of the building to be developed as the household's private realm. Once the ideal slope and access routes have been found, other "infrastructure" considerations such as sewage disposal, water, and electric service should be carefully considered.

The discovery of a ledge on the building site may at first seem to be a problem, but, in fact, can be an asset. Blasting ledge for an average-sized foundation may cost up to

$2,000 (including cleaning out the rubble) but there are advantages. In most cases the rubble is welcomed as fill for a driveway roadbed or level terrace on the downhill, sunny side of the house. Furthermore, if the footings for the (presumably) exposed south wall can be built at grade, on ledge, rather than below the frost line, there will be considerable savings in material, labor, and excavation costs. Blasting is also likely to open up seams in the ledge under the building proper, an aid to drainage if water should find its way under the building.

Ledge Can Be a Problem

Ledge *is* a problem if it covers the *entire* site. Water and utility line trenches have to be blasted, and on-site leach fields over ledge are virtually impossible.

More critically, a ledgy site will by definition have very little soil in it, and one of the essentials for the completion of an earth-sheltered building is a vast quantity of earth. An almost inestimable amount is consumed before the final topsoil is added. Between thirty and forty six-yard dump truck loads are required for the roof alone. Another couple of hundred loads can disappear in no time around the sides of the building. A minimum of 1,500 cubic yards of earth at $2 or $3 a yard is an expensive extra if it must be imported instead of found at hand on the site. Most of the cost of imported fill will reflect the distance it has to be hauled. The closer the source, the better.

If fill must be brought in, three types should be specified. Most of the material can be sandy, gravelly, even rocky rubble—cheap, bulky fill that won't hold water. For backfill-

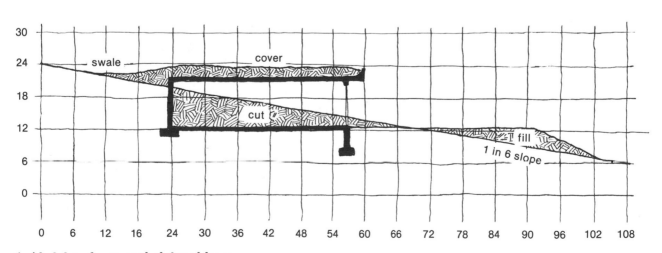

An ideal slope for an earth-sheltered house.

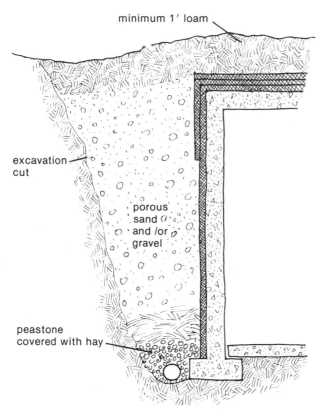

minimum 1′ loam

excavation cut

porous sand and /or gravel

peastone covered with hay

What surrounds an earth-sheltered house.

ing next to the building, a clean, porous sand or gravel, free of large rocks (over three inches), will provide a good drainage and prevent damage to the walls. Covering the fill and the roof should be a rich, organic loam, a minimum of six inches deep on the side slopes and a foot deep on the roof, sown with "conservation mix" grasses and mulched with hay immediately. Finish grades on side slopes should incline no more steeply than a one-in-two pitch.

Depth of Soil

The soil on the roof should be deep enough to maintain vegetation, but should not be much deeper for its insulation value. Earth is a poor insulator, especially when wet or frozen. The cost of the added structural bulk required to hold each additional foot of soil on the roof far exceeds the cost of an additional inch of rigid insulation. An inch of polystyrene foam (R-5.4) compares very favorably as insulation to a foot of soil (R-1.3). Unless it is extremely deep, the earth cover on the roof of an earth-sheltered building is the least important area of earth-sheltering around the building. What makes earth work as a *moderator* (not an insulator) is its integrated mass, its connection to the earth proper. When there's enough of it, it assumes a stable and relatively benign temperature. On the other hand, if we isolate it from itself, in shallow layers, it will assume the thermal characteristics of the ambience around it, the air above

Evidence of a strong structural system comforts us underground.

it, and whatever heat should escape through the insulation from the living space below.

Why put soil on the roof at all? Why not just add more insulation? There are a number of good reasons. Earth on the roof means fire protection and lower yearly insurance rates—by up to 30 percent. It's a good moderator in spring and fall when thermal cycles are sudden and extreme. It restores several thousand square feet of vegetation that is typically lost forever, and it affords rich possibilities for integrating the housing into the landscape.

Awkward Alternative

Conversely, leaving the roof exposed saddles the designer with an awkward aesthetic problem. When seven-eighths of a building is buried and the exposed eighth is the top of a wall and a roof, the potential for architectural expression is limited indeed. Equally difficult is the technical detailing against weather and foot traffic. Another important factor is the loss of at least two feet of cover above the walls which, in effect, brings the entire building two feet up out of the ground. Ground temperatures vary from ambient air temperature at grade down to a stable range at about ten feet. This range varies in the United States from north to south, 45° to 60°, and from summer to winter by about 5° with a two- to three-month lag between seasons. Pushing the building as deep into the ground as possible simply puts it in contact with earth that has more stable temperatures, warmer in the north, cooler in the south.

The elevational plan, with glazed south wall.

South

Compromise

Earth-sheltered houses can be designed in an infinite variety of configurations. The most energy-efficient have the fewest openings to the outside and are undoubtedly, from a humanistic point of view, the least gratifying in which to live. The value of natural light and ventilation can't be overemphasized. The compromise in energy lost is far overcome by the quality of life gained by direct contact with the natural world.

The simplest and most common configuration is the "elevational," or rectangular, shoe-box shape with the long axis running east and west. The north wall and roof are buried, as are parts or all of the east and west walls. The south wall is glazed for view, ventilation, and passive solar input. Primary rooms are arranged along this southern, preferred wall, with circulation behind and secondary spaces against the north (dark) wall. The "elevational" plan and its variations are limited by its linear circulation pattern which becomes monotonous and space-consuming if it must serve too many rooms as a double-loaded corridor.

Costs Increase

The multi-story elevational plan eliminates some of the linear circulation problems and is appealing for a steep site. Costs per square foot of living space will increase, however, due to additional retaining walls and structural requirements imposed by increased lateral loads attendant to greater depths into the ground. As a passive solar collector, the multi-story plan is an ideal configuration.

The "atrium" plan is attractive on sites with hostile or limited environs such as an urban ghetto or a lot stuck between a six-lane highway and a steel mill. Intrafamily privacy is compromised, however, as all the primary spaces will face into the atrium. The size of the building is limited by the size of its courtyard, and as size increases beyond three or

The atrium plan, ideal for unattractive sites.

four bedrooms, the circulation patterns around the atrium become extravagant or awkward at best. But for a small program on a small lot with no views, the atrium approach may be best.

Use of Berming

The other most common configuration is the bermed or semi-buried scheme. This is usually seen on flat sites where, because of no slope or a high water table, the building can't be sunk into the ground. Instead, the earth is bermed (sloped) up, around and over the building. As long as enough volume of earth is used, berming creates the same moderating effect around the structure as the earth against the back wall of an "elevational" plan.

Most of the materials and methods used in building earth-sheltered houses are familiar to any competent house-builder. Reinforced, poured-in-place concrete is the most common component used in footings, slabs, walls, and roofs. Precast, pretensioned concrete planks,

wooden trusses, and heavy timber are common roof systems.

Standard Footings

Footings are typically the standard sixteen-inch width, eight to twelve inches deep, poured with 2500–3000 psi concrete with ¾- to 1½-inch aggregate. A no. 4 rebar (half inch) may be placed near each bottom corner, running the length of the footing. Short pieces of rebar called dowels are left sticking up out of the footings a minimum of fifteen inches on twelve- to sixteen-inch centers. These dowels will be "tied" (wired) to the vertical rebar in the walls. In a single-level building, an eight-foot wall will typically be eight or ten inches thick with no. 4 bars running twelve to sixteen inches on center both horizontally and vertically, spaced one inch from the inside face of the wall. The concrete specified will have a minimum strength of 3000 psi and the aggregate will be ¾ inch. The thickness of the wall and the exact spac-

Roof systems on earth-sheltered structures are designed to carry loads at least four times greater than roofs on above-ground structures.

48

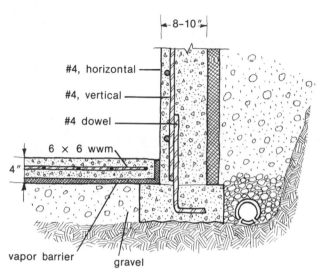

8-10″

#4, horizontal

#4, vertical

#4 dowel

6 × 6 wwm

4″

vapor barrier

gravel

Typical footing for a slab wall, strengthened by horizontal and vertical rebar.

ing of rebar will be determined by the lateral loads against the wall. Given the field conditions and a few critical dimensions, an engineer can quickly determine the exact specifications, but they will invariably be within the parameters discussed above.

Concrete walls running more than thirty-five or forty feet in length are usually broken by a control joint to allow for shrinkage during curing. The control joint is a vertical divider, usually an interlocking neoprene strip designed for this specific purpose to be flexible and watertight. Rebar extends through the joint, but the concrete is left isolated, one side from the other, so that, as it cures and shrinks, the control joint will take the place of shrinkage cracks. Even in a sixty-foot wall, shrinkage will amount to less than an eighth of an inch, but it is better to limit it to one acceptable point than anticipate a number of random hairline cracks along the wall.

Roofs

Roof systems on earth-sheltered structures are designed to carry loads at least four times greater than roofs on above-ground structures. One foot of saturated soil, deep snow cover (fifty pounds per square foot), and a pedestrian load add up to a total of 230 pounds a square foot. In order to carry these kinds of loads, we turn to concrete, lightweight steel open-web joists, and heavy timber construction.

The most widely used roof system is the

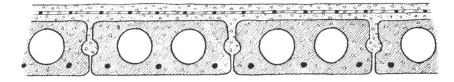

Precast, prestressed concrete plank.

precast, prestressed concrete plank system. Precast planks are cast in modular widths and lengths, trucked to the site, swung into place with a crane, and locked together with either a grouted joint or welding-plate welds. Assuming a soil cover of twelve to fifteen inches, precast plank spans are practical up to about twenty-four feet. Some planks are hollow, providing built-in conduit and duct passages. Others are "T" or double "T" or a flattened "U" shape. Some systems require a thin concrete topping while others allow waterproofing to be applied directly. Precast planks are ideal for repetitive, modular schemes but are costly if changes in the basic format are required. Angled end cuts, skylights, or mechanical equipment openings are anathema to the economy of mass-produced, precast systems. The underside of the plank is usually painted and treated as the finished ceiling, an acoustic and aesthetic compromise that many are unwilling to live with.

Concrete slabs

Poured-in-place, reinforced concrete roof slabs are more expensive than the precast plank and are limited in span (by cost) to the twenty-foot range. On the other hand, poured-in-place concrete is infinitely flexible, can assume virtually any shape, slope, or thickness, and offers a monolithic roof that can even be poured at the same time as the walls. A flat concrete ceiling is aesthetically and acoustically unappealing, but an arched, domed, or irregular concrete configuration might be pleasing.

Lightweight steel open web joists

Lightweight steel, open web (bar) joists, with steel decking and a three-inch concrete topping (or wood decking) above form a simple, economical system appropriate for up to twenty-four-foot spans and modular, rectangular designs. The joist depth increases considerably as the spans increase. A twenty-foot precast, prestressed plank may be only eight inches deep, while a twenty-foot open-web joist designed for the same loading would be close to three times that depth. This extra depth adds almost two vertical feet of unlivable space to the building's heating load. The open-web joists provide a handy route for wiring, piping, and ductwork and may require a hung ceiling. These remarks apply in a general way to wood trusses also, taking into account that the wood trusses are inherently weaker than the steel and will be even larger in width and depth when designed for the same span and load.

Heavy timber systems

Heavy timber roof systems are structurally limited to the sixteen-foot span range. In areas where native lumber is available, however, heavy timber may be the least costly, least energy-intensive system available. The wood system increases the overall roof R-factor by three or four points and is acoustically ideal and aesthetically pleasing. It is flexible and lends itself particularly well to creative expression. A sixteen-foot span with a 230-pound total roof load would require, for

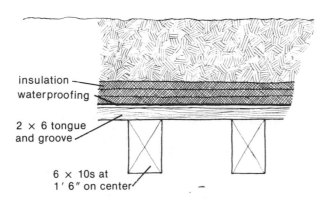

insulation
waterproofing

2 × 6 tongue
and groove

6 × 10s at
1′ 6″ on center

A heavy-timber system for a twelve-foot span.

instance, structural grade Douglas fir or yellow pine (1,500 psi) 6 x 10 timbers, spaced at sixteen inches on center. These would be decked over with 2 x 6 tongue-and-groove kiln-dried spruce or fir, leaving a beautiful, exposed ceiling that never needs to be maintained. Interior bearing walls may also be wood, such as 2 x 6s at twelve inches on center (up to eight feet tall) for the above sixteen-foot span and load.

Here are some of the advantages and disadvantages to such a roof system.

Advantages
1. Low cost. In 1981 in New Hampshire, 6 x 10 hemlock could be bought for twenty-four cents a board foot.
2. It's available in any quantity. There's no penalty for orders that would be considered small by concrete plank manufacturers.
3. No heavy equipment is needed for installation.
4. The system provides a beautiful and acoustically superior finished ceiling.
5. It reflects a warm quality of light.
6. It adds two to four points to the R-value of the roof.
7. It is easily manipulated for holes for vent stacks, chimneys, and skylights.
8. The heavy timber system is the least energy-intensive to produce, ship, and erect, and uses a renewable resource.
9. It is lighter in weight than most alternatives, and therefore is less dead-loading and can be used with lighter bearing walls.

Disadvantages
1. Termites can be a problem, *if* they get through the waterproofing or enter from the inside.
2. The span and load limitations prohibit spans much over sixteen feet with loads much over 230 pounds per square foot.

The woods most commonly used for this system, listed in their order of strength, with the strongest first, are yellow pine, Douglas fir, red or white oak, eastern hemlock or white spruce, and white pine.

Using large timbers will be much easier if you get cooperation from your sawmill, and if the sawmill has the equipment to do what you ask.

Ask the mill to dress (plane) the top surface of the timbers so as to establish a uniform dimension, top to bottom. This dimension is usually unpredictable with rough-sawn lumber.

Have the timbers cut square on one end, or both ends if you are certain of exactment measurement requirements.

And, if you use rough-sawn studs, have both edges dressed to establish a uniform dimension side to side.

Anticipate shrinkage with any heavy timber that is used green. A green 6 x 10 hemlock timber may shrink to 5⅝ by 9½. Properly anticipated, shrinkage shouldn't cause problems.

There's no need to worry about staining or sealing roof timbers and decking material. The wood will develop a natural patina and slightly darken as it weathers the years inside your home.

Insulation

Insulating the earth-sheltered building is of critical importance; outside, for the prevention of heat transfer, and inside, in some circumstances, for the prevention of condensation.

It's generally agreed that as much mass as possible—chimneys, slabs, walls, roof structure, and partitions—should be encapsulated within the building's envelope of insulation. This approach provides thermal storage capacity without the redundancy of otherwise nonfunctional elements such as tanks, rock beds, or freestanding (and usually irritatingly located) Trombe walls.

The temptation may be great to incorporate a picturesque boulder or cliff into one's living room. It will add to the room's thermal storage capacity only if it is isolated from outside temperatures. Stone is an excellent conductor as well as a good storage medium, and, if the cliff or boulder is exposed outside the house as well as in, the flow of heat through the stone will constantly conduct heat out of the building.

Any dense material such as concrete, stone, quarry tile, slate, and even wood will absorb large amounts of Btu. Wherever it can be integrated into the design, water is by far the most economical storage medium, with a Btu capacity per pound of nearly three times the capacity of concrete or stone. (Eutectic salts are the most efficient medium available, but are, to date, prohibitively costly.)

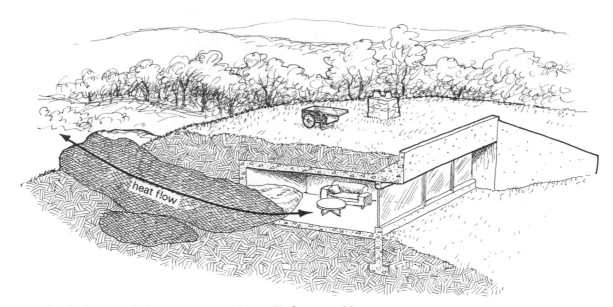

Incorporating boulder or cliff into home means heat will flow out of it.

Thermal storage mass is useful as a means of retaining excess solar input as well as off-setting sudden hot-cold cycles. Most of its value is measured in direct proportion to the amounts of solar overload the building is likely to receive. South-facing rooms make optimum "passive" use of mass, especially in areas where sunlight is radiated directly onto the mass. Unless the overheated air from these sunny front rooms is pumped elsewhere in the house, there is likely to be inadequate absorption taking place. It will be uncomfortably warm in these rooms, 80°–90° F., while the dark, rear rooms may be on the cool side. Air can be moved from these overheated rooms to the underutilized storage capacity of the walls and other mass of the rear rooms simply by thermostat-activated fans. An even distribution of excess heat will take maximum advantage of every unit of storage mass available.

Insulating the heated surface of a potential thermal storage mass may sound blasphemous, but, in moderation, it has its logic. In humid, hot weather, some closed-off rooms or corners of the house may receive little or no circulating air. The concrete walls are insulated on the outside, to be sure, but their low specific heat can bring their temperature down to the critical dew point. At this point,

vapor will condense out of the superhumid air and cause the walls to sweat and ultimately grow mildew and other micro-organisms. This will occur, if at all, in a limited number of spots. The most likely areas are in dark, poorly ventilated rooms, especially near the footings. The footing below the wall can't be insulated on its bottom surface, so it conducts the earth's relatively cool temperatures upward to the lower portion of the wall above. Condensation can be avoided altogether with the application of a half-inch layer of polystyrene board (R-2) to the interior surface of the concrete, finished off with 1 x 3 strapping, six-mil vapor barrier, and gypsum. This system will eliminate the condensation and only slightly compromise the wall's thermal storage capacity.

Below-Grade Insulation

Extruded polystyrene, a rigid, closed-cell, dense foam board, is the most extensively used below-grade insulation on the market.

Styrofoam, manufactured by Dow Chemical Co., is one of the most popular brand names. Its R-value is rated as 5.5 per inch. It's fabricated in 2 x 8-foot panels, one or two inches thick, with square or shiplapped edges. It can be used under a slab, against walls, and on top of the roof. Where more than one layer is used, joints should be overlapped to avoid direct heat-flow paths. The panels are put in place against the walls during backfilling, starting at the bottom and working up the wall. The backfill holds the insulation in place. The foot or more of soil on the roof holds the insulation in place and prevents it from floating.

Expanded polystyrene bead board is also used in these applications. It comes in different grades and with various brand names. Some of the poorer grades will absorb a certain amount of water. Cost per square foot and R-values of bead board are lower than the extruded polystyrenes, and it is compositionally less dense and therefore more vulnerable to damage during construction.

Polyurethane isocyanurate foam, notably Thermax, manufactured by the Calotex Corporation, is another below-grade insulation board. Its initial R-value of eight per inch is subject to a 20 percent loss in performance as the product ages. It is manufactured in 4 x 8-foot sheets in a variety of thicknesses, and is more costly per square foot than either of the polystyrenes.

Insulation Over Waterproofing

When the insulation is applied to the outside of the building, it's applied *over* the waterproofing membrane. Until the development of the closed-cell polystyrene and urethane foam insulations, the typical sequence was reversed. The insulation was applied first and

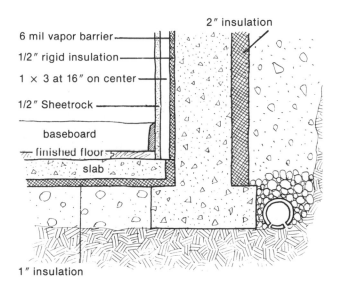

Insulating heated surface of wall makes sense in hot, humid areas.

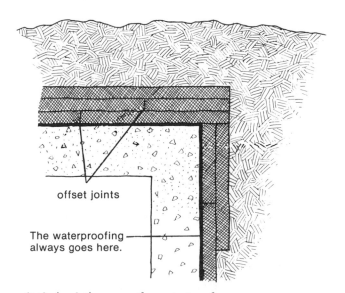

Apply insulation over the waterproofing.

then covered by the waterproofing. This led to problems that the more recent inverted systems have solved.

Insulation outside the membrane provides three important advantages:

1. Bonding the waterproofing directly to the structure makes it easy to pinpoint the source of a leak, because once it has penetrated the membrane, the water can't migrate laterally before it appears inside some distance away. In the *old* system, the insulation board below the waterproofing acted as a conduit for the lateral movement of water, making punctures difficult to locate, and repair much more costly.

2. Applying the insulation to the "hostile" side (outside) of the membrane eliminates harmful thermal cycling, keeping the membrane at a temperature as even as the inside of the building itself.

3. Loose-laid insulation board on the outside of the waterproofing is easily removed if repair is necessary, and provides an important layer of protection against punctures by stones and traffic during backfilling.

Cover Gravel Base

For insulation below the floor slab (in cold weather sites only), a compacted gravel base is first covered with a six-mil polyethylene vapor barrier. The insulation board is put down and covered with 6 x 6-inch, 10/10 woven wire mesh, then covered with four inches of 2500 psi, ¾-inch aggregate concrete.

Earth-sheltered buildings are insulated differently in different climates. In northern New England, for instance, the difference between inside room temperature, 70 ° F., and average soil temperature, 45 °, is 25 °. This

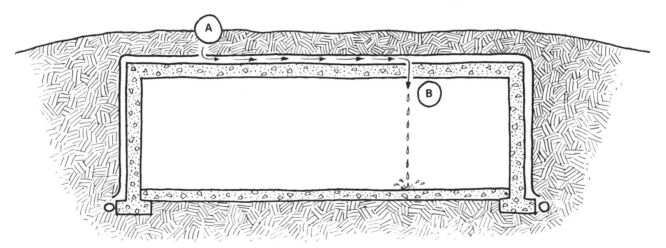

If waterproofing is not bonded to the structure, water may leak through, then travel laterally before appearing inside. This makes it difficult to find the leak.

differential in temperature is referred to as $\triangle T$. In this example, the $\triangle T$ is 25 ° and a *deficit*. This deficit is constant. The earth around the building is a thermal sponge with enormous capacity, constantly absorbing heat. In hot climates, however, this exchange is an asset and is welcomed. An unheated, earth-sheltered building with no exterior openings will eventually assume the temperature of the soil around it. An unheated earth-sheltered building with a generous amount of south-facing glazing will assume a temperature somewhere between outside air temperatures and those of the soil. But if the outside air temperature is low enough and there is no sunshine, the unheated building will freeze. The likelihood of an earth-sheltered building freezing increases in direct proportion to its non-earth-sheltered elements—walls, roof, windows, doors, skylights.

Soil Temperatures

In hot climates, average soil temperatures are higher (60 ° - 65 °). If inside room temperature is targeted again at 70 F., the differential ($\triangle T$) between inside and out can be thought of as a 10 ° *credit*. The earth is pulling heat out of the building, so insulation will be used only on the roof and the upper foot or two of the walls where soil temperatures will average *above* the desired 70 ° inside temperature. The rest of the wall and floor areas are left uninsulated so as to conduct heat away from the building into the (10 ° cooler) soil.

The earth-sheltered building tradition has been strongest in hot climates. Examples of indigenous applications of earth-sheltering principles have been common to the south

In hot areas, heat will be drawn from an uninsulated building.

Earth-sheltered buildings have been around for centuries.

Mesa Verde, where for centuries the natural properties of adobe have been used to advantage.

PHOTO BY ROBERT BENNETT

and east of the Mediterranean for centuries. In the United States, native-American use of earth-sheltered techniques was developed mostly in the Southwest where the natural properties of adobe were brilliantly exploited as a medium of thermal storage and exchange.

Waterproofing

The most worrisome component of an earth-sheltered building is the waterproofing. Above-ground buildings may leak now and then, but the assumption is that the leak can be easily found and fixed. Fixing a leak on an earth-sheltered building begins with the removal of earth and insulation, easily done if the leak is on the roof, not so easily done if it's just above the back wall footing.

Why Waterproofing Fails

Waterproofing failure on an above-grade building can invariably be traced to one or more of the following:

1. Poor method of application.
2. Inappropriate selection of roofing material.
3. Punctures, due to such things as nails, rocks, and foot traffic.
4. Flashing failure at a vent stack, chimney, or skylight.
5. Shrinking and swelling-cracking due to thermal cycling.

6. Decomposition due to ultraviolet radiation from the sun.

With care, the first four categories can be avoided. The remaining two problem areas are eliminated by covering the building with insulation and earth fill. Water itself remains the only threat, and there are a number of waterproofing products which have been proving themselves convincingly for years in underground applications.

On the other hand, waterproofing is still part science, part craft, and part experiment. Waterproofing is the Achilles heel of the underground building. Moreover, the psychological despair associated with subterranean leaks doesn't compare with the lighthearted old bucket-under-the-drip attitude.

Study Site Conditions

Waterproofing an underground building begins with an analysis of site conditions. How much water is there likely to be, and where is it likely to be coming from? Hydrostatic pressure and a high water table will require an extensive treatment on all surfaces of the building, including under the footings and slabs. On the other hand, a high and dry site will require no treatment under the building (except for the typical vapor barrier), standard treatment on the roof, and only moderate applications on the walls. If cost were no object, one would be inclined to waterproof every surface. No harm would be done, but

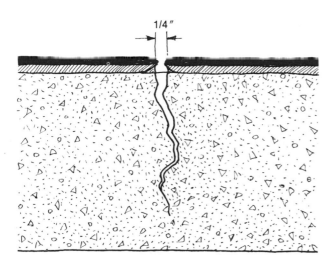

1/4″

The ideal waterproofing will be flexible enough to expand without cracking.

costs are substantial, ranging from $2 to $5 per square foot (materials and labor, 1981).

Another critical consideration is the compatibility of the waterproofing material to the structure, and the availability/reliability of the product and technical help or equipment needed for installation. For instance, a domed shell is a poor candidate for sheet-type materials, steep slopes require a firm bond between the structure and the waterproofing, and high-temperature liquids are difficult to apply to vertical surfaces.

Products Available

There are dozens of waterproofing products that claim to be suitable for below-grade applications. In making a selection, bear in mind the functions the waterproofing membrane will be required to perform.

1. It must be *waterproof*—not *vapor* proof or *damp* proof or any other euphemism.

2. It must retain its flexibility, its ability to expand and contract without cracking.

3. It must be either very tough or self-sealing in order to counteract the likelihood of the occasional puncture.

4. It must be chemically or mechanically resistant to acids and salts in the soil.

5. It must be capable of forming a monolithic, contiguous, uninterrupted envelope around the building.

6. If it should fail for some reason, it should be simple to repair and should *not* complicate the identification of the point of failure.

With these points in mind, the following products are recommended as the most proven, suitable, and practical means of waterproofing your underground building.

Liquid rubberized asphalt

Uniroyal Construction Products makes a rubberized asphalt product called Liquid Membrane 6125, marketed through American Hydrotech, Inc., 541 N. Fairbanks Ct., Chicago, IL 60611. The system requires professional application and is handled by roofing specialists around the country. (In upper New England, contact Jamieson, Inc., offices in Montpelier and Burlington, Vermont, Augusta, Maine, and Manchester, New Hampshire.)

Application procedure requires a clean, dry, frost-free, smooth surface, vertical or horizontal, curved or flat. A conditioner (primer) is sprayed on the surface and allowed to dry. Solid cakes of the rubberized asphalt are then heated to a lump-free liquid, up to about 400° F. The hot liquid is applied to the surface with a screed rake at a thickness averaging 180 mils (3/16 inch). At construction joints, cracks over 1/16 inch and transitions from deck to wall, etc., a strip of heavy rubber sheeting—Elastosheet—is sealed into the liquid rubberized asphalt as a means of reinforcement at these critical points.

The advantages of Liquid Membrane 6125 for underground applications are several. The product can be applied in cold weather, down

Liquid membrane is spread on surface by "screeding."

to 0° F., as long as the surfaces are frost-free. The material is self-sealing up to 1/16 inch, and can be repaired with added applications of heated liquid. It sets up within minutes and can be traveled on immediately thereafter. Its liquid form makes it ideal for irregularly shaped surfaces and its seamless, monolithic nature eliminates the potential for leakage that may occur between the joints of sheet materials.

The only disadvantage may be in the asphalt compounds. The petrochemical industry standards for asphalt have been steadily declining since the oil embargo. Asphalt-tar products have become more brittle and prone to drying out. The rubberizing of asphalt was developed as a means of improving flexibility, and appears to be effective. As these products evolve, we will undoubtedly see the rubber components increase as the industry distills off more of the valuable oils for other uses. The Uniroyal system costs a little over $2 per square foot and up, depending upon flashing and edge detailing, and has a two-year warranty.

Rubber sheet materials

Carlisle Tire & Rubber Co., Box 99, Carlisle, PA 17013, has an extensive catalog of products and components for waterproofing underground buildings. It lists four systems in its Sure-Seal line of waterproofing systems, two of which are most appropriate for underground installations. The product names are Sure-Seal Butyl Rubber Membrane and Sure-Seal EPDM Rubber Membrane. Both the Butyl and the EPDM are manufactured in standard thicknesses of .030, .045, .060, and .090. A minimum thickness of .045 is recommended for below-grade use. Widths are standard at 54 inches and 10, 20, and 40 feet, shipped in rolls 100 feet long. Prices (from factory, 1980) for the .045 thickness are 68¢ per square foot for the Butyl and 62¢ per square foot for the EPDM. The rolled rubber sheet is very heavy and a crane is needed to move the material. Special widths or lengths, factory-produced accessories such as pipe flanges, inside or outside premolded corner-flashings, and number of splices required all increase installed costs. Technical representatives are available for jobsite assistance on a per-diem basis.

Installation of the Butyl or EPDM (Ethylene, Propylene, Diene, Monomer) is simple and straightforward, requiring more care than skill, plenty of both being ideal. Surfaces should be clean, smooth, and dry. The sheet material is rolled out into its final position all over the roof, with joints overlapping three inches. (The fewer the joints, the less labor and less likelihood of joint-related leaks.) The sheets should be allowed to relax in their final positions, with no stretching or rumpling. If it's windy, weight the edges down with 2 x 10s weighted with cement blocks for ballast.

Once the sheets are in place and relaxed, the joints are bonded together with Carlisle splice cement ($12 per gallon covers both surfaces of 140 lineal feet of splice) and sealed with a gum tape and lap sealant ($25–30 per 100 feet of splice). Prefabricated pipe flanges

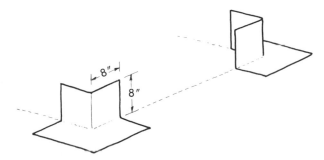

Carlisle offers premoulded corners for skylights and curbs.

and premoulded corners ($10–15 apiece) are cemented onto the sheet material in the same way the sheets are jointed, and a Sure-Seal Curing, Pourable Sealer ($16.50 per gallon) can be poured around irregular-shaped penetrations and covered at the joint with the lap sealant. Edges of the sheets terminating against parapets or down over the roof edges onto side walls can be sealed into typical reglets or mechanically terminated with Carlisle's "U-9D" system. The U-9D is simply a metal strip bolted or ramset through the sheet into the substrate, running the length of the edge. Carlisle Water Cut-Off Mastic ($15 per gallon) is caulked against the outside of the joint.

Membrane terminations and vertically applied sheets will require a bonding adhesive ($15 per gallon, covers approximately sixty square feet) to attach the material to the substrate. Vertical surfaces are typically spot tacked with the adhesive, which must be applied to both mating surfaces and allowed to

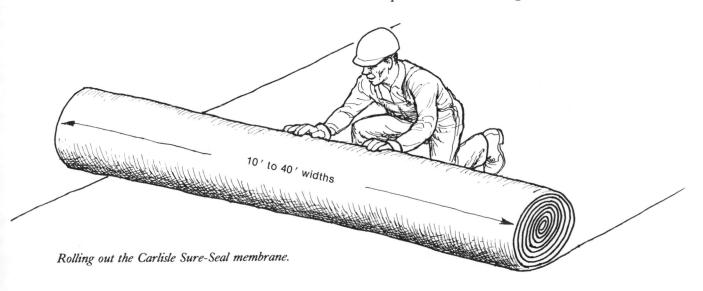

10' to 40' widths

Rolling out the Carlisle Sure-Seal membrane.

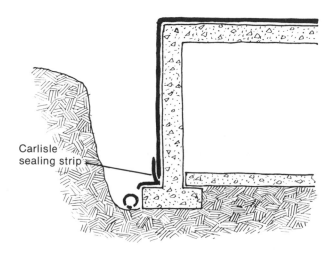

Carlisle
sealing strip

*The sealing strip holds down edges of the membrane
sheets.*

dry tack-free before bonding. Terminal edges are bonded at least twelve inches in from the edge with either the bonding adhesive or a twelve-inch-wide Carlisle sealing strip, #90-8-30A.

Bonding a waterproofing membrane to its substrate is considered good practice in any waterproofing situation. Bonding doesn't necessarily make the system more waterproof, but if a leak develops, bonding prevents lateral migration between the membrane and substrate, making the point of repair more easily identified. The Carlisle sheet material *can* be bonded (but usually isn't). Both the substrate and the underside of the sheets must be coated with bonding adhesive ($15 per gallon, covers about sixty square feet) and allowed to dry tack-free. Then the membrane must be repositioned in its "relaxed" mode, taking care not to trap air bubbles and to avoid stretching. The Carlisle materials *must* be bonded to some extent on sloping or vertical surfaces, but the labor costs are high enough to make it of questionable value on flat surfaces. The Uniroyal liquid membrane and the next two systems to be discussed are all automatically bonded systems, but even when applied loose-laid (non-bonded), the Carlisle Butyl and EPDM systems are extremely well-considered within the profession. Materials and labor combined will bring the cost of the Carlisle system up to somewhere between $3 and $5 per square foot.

In all of the systems discussed, hidden costs in flashings and gravel stops, flanges and irregular configurations will add considerably to

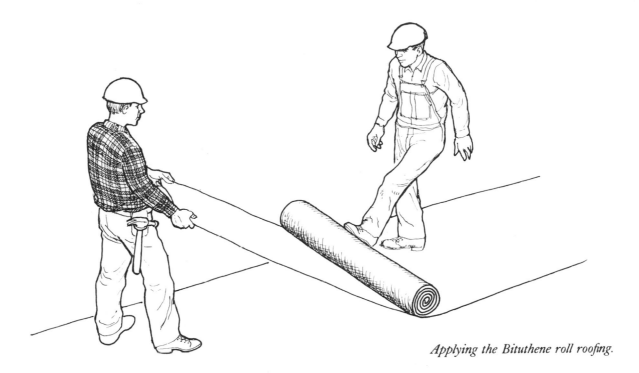

Applying the Bituthene roll roofing.

the overall square foot figure. In general, a large, rectangular roof with no interruptions will be the least expensive to cover. Economy will suffer as size is reduced and irregularities increased.

Rubberized asphalt membrane

Grace Construction Products, 67 Whittemore Ave., Cambridge, MA 02140, manufactures a product called Bituthene. Bituthene is a tough, pliable, waterproof polyethylene sheet, coated on one side with a thick factory-controlled layer of adhesive-consistency rubberized asphalt. It's manufactured in rolls thirty-six inches wide and sixty feet long, with a release paper interwound to prevent the material from sticking to itself.

There are two basic Bituthene systems applicable to underground structures. Bituthene 3000 and its companion primer must be applied at air surface temperatures above 45° F. The primer is not sticky, nor does it serve as a contact cement. Its only purpose is to fill in miniature substrate irregularities and create a uniform, dust-free surface for the membrane adhesion. It's applied with a lamb's wool roller and takes thirty to sixty minutes to dry.

Bituthene 3100 LT (Low Temperature) membrane can be applied in temperatures down to 25° F. or up to 75° F. The 3100 system is essentially the same as the 3000 except for its greater adhesive capabilities in cold weather applications. The 3100 uses the 3000-type primer on horizontal surfaces and a special Low Temperature Primer on vertical surfaces. The Low Temperature Primer remains tacky after drying, and should be covered within eight hours. It cannot be used in temperatures below 25° F., and, like all parts of the system, must be applied to a clean, dry, smooth, frost-free substrate.

Application of both the 3000 and 3100 membranes is a simple operation, again requiring more care than skill. Beginning at an edge, the first roll is rolled across the roof with one person pushing the roll from behind, and another person pulling away the release paper from in front. When the end of each pass is reached, the membrane is cut with a utility knife and lapped down over the previously prepared side wall. Each successive pass is made with a minimum three-inch overlap. (An overlap of more than eighteen inches, creating a two-ply system, is recom-

63

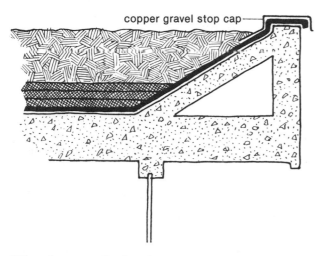

copper gravel stop cap

When sheets or rolls of roofing waterproofing membrane are used, edges are finished this way.

mended for the ultimate in protection.) Because it is self-adhering, Bituthene requires no extra adhesive at the lapped joints. End joints and special conditions around skylights and chimneys should be coated with Bituthene mastic, available in tubes or five-gallon pails.

As it's being installed, the Bituthene should be thoroughly rolled with a thirty-inch lawn-type roller to assure a tight bond between layers. "Fish mouths" must be slit, and the flaps overlapped and repaired with a patch. Edges can be sealed with the typical reglet detail or with a trowelled bead of Bituthene mastic. The only difficult part of the job is aligning the roll at the beginning of each run so that it rolls out parallel to the adjacent run. This skill is soon perfected, however, and at worst, the errantly directed roll can be stopped and cut and realigned.

Bituthene can be purchased directly from innumerable building product retailers and installed by careful amateurs or professionals. The 3000 membrane costs about 50¢ a square foot, the 3100 about 70¢ (1981). A double-lapped system, including materials and labor, will cost a little under $2 per square foot. Because it's a sheet-type material, Bituthene, like the Carlisle Butyl and EPDM, is not applicable to compound-curved surfaces. Technical advice and assistance are available through the Grace Construction Products office in Cambridge, Massachusetts.

Water-activated clay waterproofing—bentonite

Bentonite is a volcanic ash/clay composed

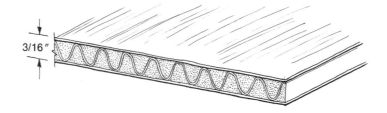

Cross-section of a Volclay panel.

3/16″

of superfine particles which expand over fifteen times their original size when wet. Bentonite is inert, non-deteriorating, and capable of an infinite number of wet-dry cycles. Once it has been refined into its pure state, it's marketed in two forms, both of which have been used extensively and successfully in the waterproofing of earth-sheltered buildings.

The American Colloid Co., 5100 Suffield Ct., Skokie, IL 60076, manufactures the Volclay Panel system and calls it "permanent and self-healing." The Volclay Panel is a 4′ x 4′ x ³/₁₆″ piece of biodegradable, kraft, corrugated cardboard. Each square foot of panel holds about one pound of sodium bentonite in the flutes of the cardboard. Once in place, vertically or horizontally, the bentonite needs only water for it to turn to a watertight paste. Bentonite expands to fill gaps, tighten around nail holes or punctures, and self-heal any penetrations up to ³/₁₆ inch wide. The cardboard decomposes within a few months, leaving a layer of clay against the walls and roof. Volclay Joint Seal, a bentonite gel, is available in five-gallon cans for sealing construction joints and tie-holes, or as a mastic for holding the panels onto vertical surfaces. Volclay Hydrobar tubes or Joint-Paks are provided for termination joints at the footings and wall/slab intersections.

Installation of the Volclay Panels is very simple. The substrate should be level, holes filled with trowelled-on Joint Seal (type P for temperatures above freezing, type G for below freezing). Panels are then laid flat or tacked into place (nails or gel), overlapping 1½ inches according to blue marker lines on the panels.

There are only two slight limitations to the Volclay Panel system. Since the panels are planar, they must be laid against flat or single-curved surfaces (i.e. all structures except compound-curved shapes).

Since the clay is essentially loose against the building after the cardboard decomposes, erosion could be a concern. But once the backfill has settled firmly against the structure, the lateral movement of water around the clay particles will be adequately limited by earth pressure (typically translated through intervening layers of rigid insulation board). The bentonite may be subject to erosion only during the first year after the building's completion. Settling and shifting of soil around the structure during this period may conceivably produce cavities through which water would have an opportunity to run and erode the clay. When moist, the clay is gelatinous and fairly firm, so that it *would* take large amounts of water at considerable velocity to erode it. Care in backfilling and water runoff patterns is called for to insure a perfect job.

Another form of bentonite is produced by Effective Building Products, 2950 Metro Drive, Suite 305, Minneapolis, MN 55420. This is a paste form of bentonite called Bentonize which can be applied by trowel or by spray gun. It can be used in every type of

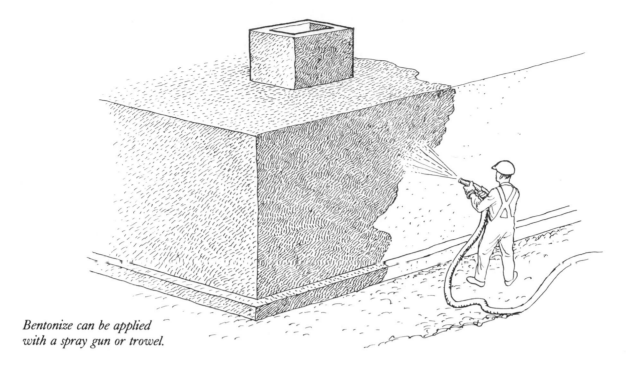

Bentonize can be applied with a spray gun or trowel.

horizontal, vertical, or compound-curved surface, is applied to about ⅜ inch thickness and sets up to a hard finish. Prices and application specifications are available from the Minneapolis office.

The Bentonize system is one of the most comprehensive and respected in the field. Except for applications over wood decks, it's suitable for every conceivable underground waterproofing requirement. Costs for labor and material are about $1.50 per square foot. The products are available through Brock-White Distributors, Minneapolis, MN.

Where to Waterproof

What about waterproofing the walls and under the slab as well as the roof? What are the most critical areas? Unless you're building in the Sahara Desert, the roof and walls will definitely need waterproofing. Building an earth-sheltered home on a site with a high water table should be avoided at all costs, but if it's inevitable, the slab will need waterproofing. (This is achieved by pouring a preliminary set of footings and slab, waterproofing over them, and then pouring a final set of footings and slab. A waterproofed slab system

may add 15 percent to the cost of a single-story building.)

Waterproofing walls

What about a well-drained site with no chance of hydrostatic pressure buildup? Here the roof would be waterproofed with one of the materials discussed above, but the walls may require a less expensive treatment. From the footings to the roof, moisture will exert little pressure against the walls as it seeps down through the porous backfill towards the footing drains. There are several economical systems of lightly waterproofing these walls. In order of least cost (and efficacy) they are:

1. Standard black asphalt foundation coating, brushed on, two coats better than one.

2. One coat of the above and one coat of plastic asphalt roofing cement, trowelled on and immediately covered with black, six-mil polyethylene.

3. Trowel on a layer of dense water-stop cement and cover with system 1 or 2.

Why they don't leak

In general, the brushed-on asphalt is a token measure. The fact that many basements "waterproofed" with this method don't leak is due to the absence of water and a good, tight concrete mix rather than the performance of the asphalt coating. Plastic cement with the polyethylene film protecting it from soil acids and salts is a much more dependable method, is economical, and no special skill or trades are required to apply it.

The intersection of the wall waterproofing and roof waterproofing will typically occur six to twelve inches below the bottom face of the roof structure. The wall treatment will have been completed first, and the roofing will overlap eight to twelve inches. It's important to determine that the two systems are chemically and mechanically compatible. Most combinations of the systems mentioned can be easily interchanged. (See diagram.)

Pitching the roof for waterproofing reasons shouldn't be necessary if the membrane is indeed waterproof. There *is* an argument for pitching a roof, however, and the pitch can be

inboard, towards interior drains. These drains will collect excess rain water and carry it down and into a runout or dry well away from the building. The popular front-to-back roof pitch is a sensible configuration for collecting sunlight, but it does direct a great amount of water to the back wall and footings, increasing the potential for water problems in these sensitive areas.

Drainage

No matter how good the waterproofing, keeping water away from the building wherever possible is the first rule of earth-sheltered design. Once the building is waterproofed, footing drains will be called upon to disperse whatever water the building sheds. Typical four-inch perforated PVC (poly vinyl chloride) solid or flexible pipe is used in exactly the same configuration as in most ordi-

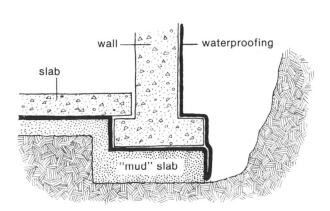

When faced with high water table, builder should pour a set of footings and a slab, waterproof them, then pour a final set of footings and slab.

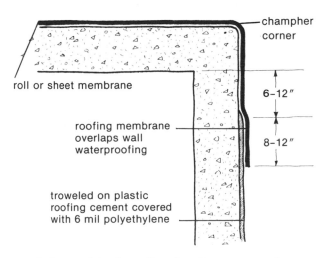

Here's how to join the wall and roofing waterproofing.

nary buildings. The pipe is placed next to the footing, under a foot-deep bed of 1½-inch stone, then covered with hay or building paper. The pipe should be pitched towards a run-out at grade, downhill from the building. (The holes in the solid pipe should be aligned at the five and seven o'clock positions. The flexible pipe has holes all around.) If erosion or freezing is likely at the run-out, a dry well or leach bed will solve the problem. Four or five cubic yards of 1½-inch stone buried one foot below grade will serve as a trouble-free collector indefinitely.

Build up base

If, during excavation, the site appears wet, i.e., standing or running water, extra stone and gravel should be used below the slab. On a dry site, an eight-inch bed of compacted bank run gravel is considered minimum good practice. In a wet situation, use an extra eight to twelve inches of 1½-inch stone below the gravel and cross it every eight feet with parallel lines of four-inch perforated PVC underdrains. The underdrains should be pitched from the upslope end towards the same run-out or dry well used for the footing drains. (The more water, the bigger the dry well.)

Mechanicals

The mechanical systems in an earth-sheltered house are similar to most slab-on-grade buildings. Cast iron waste lines and copper tubing water lines are run below the slab where required. PVC conduit below the slab simplifies the wiring routes from the panel to key locations. Since walls and roof members are often exposed or monolithically constructed, wiring, piping, and duct work are not so easily accommodated as they are in balloon frame structures.

Aside from its energy efficiency in heating and cooling, the essential peculiarity in earth-sheltered buildings is the low rate of air infiltration. The one or two air changes per hour typical to a woodframe house are reduced by 50 percent or more in a below-grade building. Not only does less outside air infiltrate inward, but stale, moist, inside air moves at an equally slow rate outward. An earth-sheltered building just doesn't "breathe" as well as its above-ground neighbor.

An elevational plan with generous amounts of glazing on the sunny sides can provide adequate natural ventilation year round. A totally buried home will need mechanical air-handling equipment (simple and common enough). In the interest of conservation, a heat exchanger can be used to temper incoming air with otherwise wasted heat from the exiting air. The Japanese firm Mitsubishi, for instance, manufactures a unit called the Lossnay which reclaims 75 percent of the heat passing through it as it discards moist, stale air in exchange for fresh outside air. Enercon Industries in Regina, Saskatchewan, also makes an air-to-air heat exchanger for under $1,000 (1981).

Outside air supply for fireplaces, wood

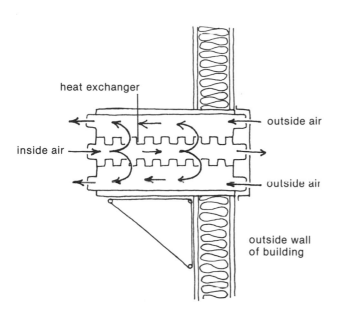

Use of a heat exchanger means less heat loss for ventilation.

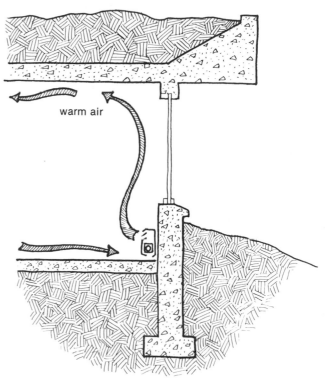

Here's a good location for a heating element.

stoves, and oil/gas burners is mandatory by code in some communities. Rather than consuming warmed air in the firing process, outside (cold) air is introduced directly to the combustion area by means of a duct (six-inch average). The outer end of the duct should be protected from wind and water with a baffle or flap that can be firmly closed when the duct is not in use.

Air Patterns

The distribution of heat in the earth-sheltered house should be designed with an emphasis on air patterns moving up the outside walls and windows and then inward. The heating medium (air or hot water) will be wrapped around the perimeter of the building, concentrated at points of maximum heat loss such as windows and doors. A wood stove in the center of the building may radiate heat intensely for a limited area around it, but a fan may be required to distribute warmed air to the remote corners of the house.

Central heating systems are common in earth-sheltered houses: wood, coal, gas, oil, or combination fuel. Forced hot air and forced hot water deliver the heat. Forced hot air systems are able to humidify, dehumidify, filter, and cool air as well as heat it—all through the same system of ductwork. Because the ductwork must otherwise be insulated and routed below the slab, hot air systems help justify the double-floor storage-distribution system popular with many contemporary designers.

4

DOUBLE ENVELOPE

The great American dream was rudely interrupted in 1973 with the formation of the OPEC alliance. The petroleum that fueled our fast-car-freeway fantasy was suddenly unavailable at the good old giveaway prices as the myth of infinite economic growth collided with the reality of finite fossil fuel reserves. Modern America's four-barreled dreamboat was running out of gas.

In a free enterprise system, a change in the marketplace rewards the nimble entrepreneur who can adapt to evolving needs. The multi-billion-dollar corporate giants, however, can only respond awkwardly to unanticipated change. Detroit automakers are a good example. Quantum paralysis stuns the giant. While the Japanese imports stole the market from under its nose, Chrysler announced its first line of fuel-efficient cars—seven years late.

A Flexible Industry

Despite its enormous volume of business, the housing industry in the United States is remarkably decentralized, flexible, and responsive to market conditions. There are no cartels, no monopolies. Most housing is still stick-built, unit by unit. Individuals call the shots, and the wisdom of their choices is quickly enough verified. Market momentum, in any field, is difficult to overcome—even *more* difficult when the manufacturers are huge and crippled with inertial capital investments predicated upon obsolete premises. The house-building business is an example of free enterprise at its best. The tool-and-die commitments that plague the heavy industry sector are absent. Competition demands adaptability and adaptability means survival.

When the first rays of the Solar Age permeated consumer consciousness, the home builder responded eagerly. Subdivisions were surveyed with southern exposure in mind. R-factors became part of the sales pitch. Double-glazing, heat pumps, greenhouses, and chimneys for woodstoves became sought-after accessories that could mean the difference between money down and thumbs down. While General Motors was still peddling the comforts of the Big Car, energy-efficient houses were being produced by the hundreds of thousands by Tom, Dick, and Harry construction companies all over the country. Experimental work proliferated by the back roads as grass-roots tinkerers looked for a better way to trap a Btu.

Purity of thought is the province of the individual. The anarchist-artist likes to trot out the old definition of the camel as the horse designed by a committee, a fair enough observation when applied to a single work of art. But collaborative design, especially in research, is a social rather than creative notion, and the fruits of its labor are convincing. In house design, the camel-horse metaphor becomes relevant only after the building crosses the line between shelter and architecture, between a *type* of house and a *specific* house.

In the development of generic house types, ideas evolve through the contributions of many. There's no patentable idea, and fortunately very little proprietary selfishness about

The Burns double-shell house was designed and built by Community Builders of Canterbury, NH.

"original" work. The double-envelope concept is a good example of collaborative evolution.

Tom Smith, Lee Porter Butler, Community Builders (Don Booth *et al.*), and Hank Huber are among the visible, vocal pioneers of the concept. Beginning with Smith's California house in 1977, the double-envelope theory proposes an intriguing approach to energy efficient design.

Sunspace

Three basic components distinguish a double-envelope house from an "ordinary" house. First of all, the south side of the house is given over to a *sunspace*. The sunspace is usually the full width (east to west) of the house and is often the full height. The south (outside) wall of the sunspace is glazed, the east and west (end) walls are fully insulated, and the north (inside) wall is also insulated and equipped with operable windows and doors. This wall is identical to the one you would find on the house if the sunspace were not included.

The sunspace is typically from six to twelve feet wide and is used for pedestrian circulation, growing plants, and intermittent living space. The south windows of the home proper look out into the sunspace which is, in effect, a huge walk-in solar collector.

Convective Loop Plenum

The second component is the *convective loop plenum* (air passage) above the ceiling and within the back wall. The plenum provides an avenue of air circulation leading from the top of the sunspace across the top of the

Cross-section of a typical double-envelope house.

house and down the back wall to a crawl space or basement. The plenum runs the full width of the sunspace (east to west) and is typically eight to twelve inches deep. At the roof line and north wall, the plenum divides the outer layer of the building into two layers, like two pairs of socks with an air space between. Both layers, inner and outer roof and walls, are insulated (see diagram), and the air moving between them is circulated, mechanically or by natural convection, around and around the "inner" house, providing an aerodynamic bath, enveloping the shell-within-the-shell with warmer-than-outside air.

Crawl Space

The third component of the double-envelope design is the *crawl space* or *basement*. Its purpose is threefold. It acts as the final link in the air circulation loop around the house, it serves as a heat and moisture storage bank, and it provides a low-grade, geothermal boost to the air buffer between the envelopes.

The floor of the crawl space is typically dirt or sand, occasionally a concrete slab. The exposed earth regulates humidity as it absorbs and disperses moisture from the air circulating over it. Although the heat storage capacity of the earth is calculable, it was greatly overestimated in the original conceptual explanations of the double-envelope design.

Why It Works

What *really* makes the double-envelope design so effective? We know it does work. Energy consumption is extremely low. Objectively monitored double-envelope houses perform almost miraculously. But how?

Madison Avenue has taught us that in order to sell something, you have to have an angle. Toothpaste assures the user of a Hollywood sex life. Soda pop makes us feel "free." A certain cigarette transforms local weather into eternal spring, and a dozen forms of aspirin are, indisputably, the only cure available for the incurable common cold.

Those Arrows

The double-envelope design lends itself nicely to Madison Avenue techniques. Its promotional gimmick is too graphic to resist: a cross section shows a house within a house. In the air space between the envelopes are big arrows. Big red arrows indicate the direction of hot air flow. Big blue arrows show which way the cold air travels.

Scene one shows midday sunlight warming the sunspace, pushing air up into and across the attic plenum. (Red arrows.) As the warm air moves along its circuit, it loses heat to both inside and outside walls of the plenum, and the arrows turn reddish blue. Down the back wall and into the crawl space, losing heat . . . arrows turn almost entirely blue. Moving across the crawl space, southward, towards the bottom of the sunspace, arrows are blue. Then they move up through the grilles, into the sunspace and . . . ahaa! Blue arrows turn to red again! The cycle is completed and a dozen big red and blue arrows tell us how neatly it all works. Hot to cold, cold to hot.

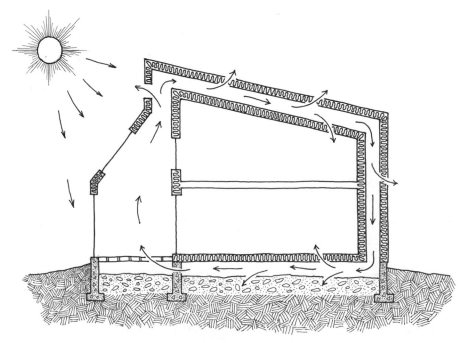

Scene one: the arrows are going clockwise.

More Arrows

Scene two: at night the sunspace gets cold (low forties). Blue arrows point downward. Cold air drainage off the glass starts a current flowing in the opposite direction from the daylight mode. Blue arrows point down through the grilles and northward into the crawl space. But what's this? The blue arrows are assuming a reddish tinge! The heat absorbed into the crawl space earth surface during the day is being retrieved back into the passing current of air! By the time the arrow has moved all the way to the north of the crawl space it has turned reddish. Up into the north wall plenum and the arrow is more red

than blue—the air is being warmed by heat stored in the plenum walls. Red arrows turn the corner at the top of the north wall plenum and point southward towards the top of the sunspace. As the warm air reaches the night-cooled sunspace it is siphoned down the cold outer glass wall and the circuit is ready to begin again with a big blue arrow.

If we are paying close attention to the implications of the red to blue arrows, we realize that in order for convection to work, a change in temperature must occur. During sunny hours the attic and north wall plenum are counted on to produce this change—a *drop* in temperature. This drop is a result of heat *loss* through the outer envelope. This aspect of the convective loop seems a bit per-

Scene two: they reverse without heat input from the sun.

A double-envelope house is virtually a house built within a house.

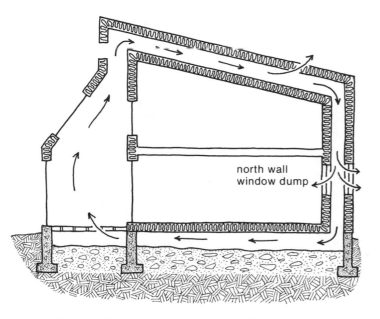

north wall
window dump →

North-wall windows can be used to dump heat.

verse. Some designs include windows in the north wall to further increase the differential by "dumping" heat out through the glazing.

But the paradigm is so simplistic, so reductive, and yet so apparently *sensible* that we can't resist. It's a picturesque demonstration of convection. Warm air rises, cold air falls. The hot-cold differential between sunspace and north wall, between crawl space and attic plenum sets up the requisite dynamic to push-pull air around the loop. No fans, no noise, no mechanical assistance. Unlike the earth-sheltered or superinsulated houses, the double-envelope can be portrayed as a kinetic system—a living house encircled by hot and cold arrows.

In a shameless excess of hyperbole, one proponent of the double-envelope woos prospective buyers with the following flights into fantasy: "The air in your home reaches you with a refreshing tang of richness. You notice and are pleased at the delightful hint of fresh strawberries and mint that float in the air. . . . Into the shower. Ahhh. The streams of hot tingling water do feel good this morning. Is it just your imagination or does solar-heated water actually feel better on the skin? . . . For breakfast, you consider the lush stalk of bananas and the plump ripe strawberries. Or perhaps, you'll wish to start with fresh-squeezed orange juice prepared from the oranges you have grown in your greenhouse. The aroma of the sweet fruit and vegetation stirs your memories of all those distant tropical islands you have longed to revisit. . . . The air temperature and humidity have pleased your

Ficus benjamina. It has never looked so happy. . . ."

Head Start

Any sober assessment of double-envelope design dynamics comes up with precious little information as to the tactile quality of solar-heated hot water. It does support, however, the bottom-line claims of energy efficiency, and the secret is not to be found in big blue and red arrows and the smiling Ficus benjamina. Before a single breath of air is moved around the convective loop, the house has an enormous head start on its standard competitors.

To begin with, a double-envelope house is virtually a house built within a house, *both* of them insulated to equal or exceed standard single-wall construction specifications. Double- or triple-thick insulation in *any* type of building will enhance its performance enormously. Secondly, the design typically arranges openings so that there are no net losses through glazed areas. A due south window is acknowledged to produce a net heat gain in most temperate climates in the northern hemisphere. The double-envelope configuration opens most of its south face to the sun and often has no glazed areas on the east, west, and north. This is a sound and essential principle of solar-heat-assisted buildings of all types.

The sunspace is, of course, a giant hot air collector. Attached to any building, a south-facing sunspace will contribute plentifully to heat gain. The radiated air could be moved into the house through an opened window, into a storage system (rock, water, or phase-change salts) with a fan and ducts, or absorbed into a storage system in the sunspace itself. In any of these examples the sunspace bonus is an established asset before the aerodynamic convection loop is even considered.

Value of Crawl Space

The least understood component of the double-envelope house is the crawl space. Whether it is exposed earth or a concrete slab, the crawl space floor is most commonly considered to be a heat storage-heat retrieval mass which transfers Btu to and from the air moving across it. Monitored vertically at six-inch increments, the crawl space floor shows only a slight variation from ambient earth temperatures. The top few inches will show some thermal gain or loss, but the figures stabilize as the depth increases beyond eighteen inches. A typical variation in air temperature, earth surface temperature, and earth temperature six inches below grade shows the earth surface temperature holding within 2° or 3° F. either side of 52°. Air temperature is 2° or 3° above that during the day, the same amount below at night. Six inches down into the earth the readings are typically a few degrees cooler than the surface during periods of insolation and equally warmer at night or on cloudy days. The six-inch depth temperatures are more stable than the surface and air temperatures above. As the depth into the earth increases, temperatures are even less responsive to changes from above and soon approach the ambient earth range with only a few degrees' variation from season to season.

The crawl space data are consistent with what our common sense tells us. The earth affords thermal storage potential to the heat source (air) above it, but it absorbs very little because:

1. The air-to-earth temperature differential is slight—20 ° F. or less.
2. The air moves across it at a slow rate and therefore exposes the earth to few Btu.
3. In convected form (as opposed to radiated), heated air will respond to gravity and tend to rise rather than fall. In fact, by the very example of the big blue and red arrows, we know that warm air rises and will tend to stack up against the crawl space ceiling rather than force its way into the ground.

The thermal tempering dynamic of the earth-coupled crawl space is one of the unique aspects of the double-envelope design. We know that the earth around a building is a giant heat sink. Its thermal characteristics are moderate but stable. Rather than attempting to isolate (insulate) the crawl space from geothermal influence, the envelope concept makes use of it. Ambient earth temperatures in the northern United States range from the low 40s to the low 50s F. These temperatures are considered uncomfortably cool for habitable spaces. In earth-sheltered design, where most of the building is in direct contact with the earth, failure to surround the structure with insulation would be a serious mistake. If inside temperatures are targeted at 65 ° F., there will always be a 15 ° to 20 ° deficit between the living room and the backfill beyond the walls. The deficit is slight, to be sure, but it is constant. (If, on the other hand, inside

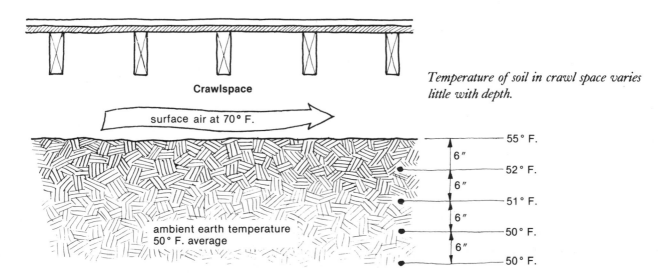

Temperature of soil in crawl space varies little with depth.

temperatures were meant to be *less* than the ambient earth temperature, the differential between inside and out is an asset.)

The double envelope provides an opportunity, however costly, to use low-grade ambient earth temperatures as a thermal boost. The ΔT (temperature differential between inside and outside air) is in effect split by the north wall air plenum. The insulated walls on each side of the plenum are heat sinks to the geothermally tempered air. This air is too cold to use inside the house, but, in cold weather, it is warmer than trapped air would be between the two insulated envelope walls. Here we have the use of earth heat as a usable resource rather than just a benign moderator and microclimate around a building.

As outside temperatures drop, the earth-tempered air in the north wall plenum becomes increasingly useful. When the weather warms up much above freezing, however, the earth-tempered air becomes a negligible asset. At 55° F. outside temperature, the 50°± envelope air is actually a thermal liability.

The crawl space ceiling (living space floor) is always insulated. Not only is the crawl space usually colder than ambient living space, but the attempt to charge the earth with solar-radiated air during the sunny day mode would be compromised if heat were absorbed up through the floor instead of down into the earth at these crucial times. "These crucial times" are few and far between, however. Figures suggest that less than 10 percent of passively convected energy ever gets stored in the earth in the crawl space.

Attic Plenum

Next to the dubious efficacy of the crawl space as a storage bank, the attic plenum is the least convincing component of the double-

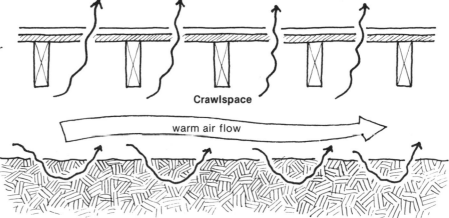

Warm air tends to rise through floor rather than sinking into soil.

Crawlspace

warm air flow

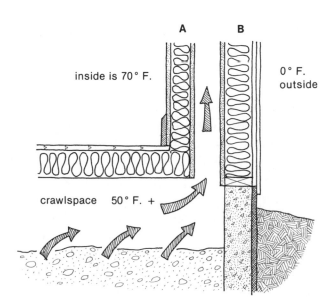

inside is 70° F.

A   B

0° F.
outside

crawlspace   50° F. +

The north wall plenum is useful only when the average of the inside and outside temperatures is colder than the crawl space air. When outside temperatures rise much above 30° F., walls A and B would be more effective if consolidated into a single, extremely thick wall.

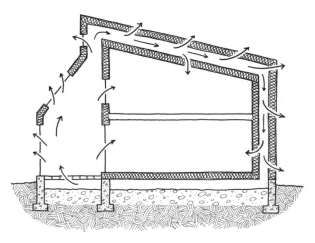

The plenum loses two to three times as much heat to the outside as it puts into the inside of the house.

envelope anatomy. While it may assist in the circulation (on cold cloudy days or freezing nights) of the geothermal boost emanating from the crawl space earth, it is also filled with 100°+ F. air during periods of sunshine. There is typically a 15°–25° F. drop in temperature between air entering the southern opening to the attic plenum and air at the point where it drops into the crawl space at the bottom of the north wall. Most of the drop in temperature is accounted for by heat loss through the outer roof shell above the plenum. Some is also lost through the outer north wall, but the roof accounts for the considerably greater share.

It seems logical—and calculations confirm—that the heat transfer from the attic envelope is moving in two directions, inward and outward. The outbound loss exceeds the inbound gain by a factor of two to three depending on conditions. This loss/gain transfer accounts for the 15–25° F. drop as the air moves across the attic and down the back wall to the crawl space. For every half to one-third Btu saved (into the house) we've lost double or triple that amount to the great outdoors.

Adding insult to injury, this scenario happens most critically during the sunniest days when internal gain is least important and potential solar gain should be put to maximum use, that is, stored away for a cloudy day. The insulated ceiling below the attic plenum would, on these days, be far better utilized if it were added to the roof insulation above the plenum and thus lessen the transfer to the outside.

Other Possibilities

We want that convective loop to work, don't we? Those arrows look so reasonable and double-envelope houses perform so well. But after we've accounted for the major advantages due to superinsulation, sunspace-collector, south-facing glazing, and geothermal input, it could be argued that there are better ways to combine these assets than to connect them via the convective loop plenums.

Eliminate Plenum?

Some of the arguments appear uncontestable. In an article in the November 1980 issue of *Solar Age Magazine,* Vic Reno suggests the elimination of the attic plenum, and the use of electric blowers and ducts for the transfer of air to storage. He also neatly dispatches the myth of inner wall surface com-

fort with a comparison of a monolithic R-30 wall and a double-envelope wall composed of two R-15 envelopes divided by an air plenum. Assuming a 70 ° F. inside temperature, 0 ° F. outside temperature, and a 50 ° F. air plenum recording, the temperatures of the living space wall surfaces were 68.5 ° F. and 69.2 ° F. respectively. The .7 ° F. advantage shown by the double wall is insignificant. Inner wall surface comfort is simply a non-issue.

Another curious paradox is the east and west wall construction. Because they are not included in the north-south loop, they are designed as monolithic walls with insulation values anywhere from R-24 up. But they are not enveloped walls and they account for over half the vertical exterior wall space in the building. The designers of the double envelope have planned these end walls to resist overall heat transfer as well as any other part of the building. If they work at the east and west, why not at the north too? Why not pump hot air from the top of the sunspace

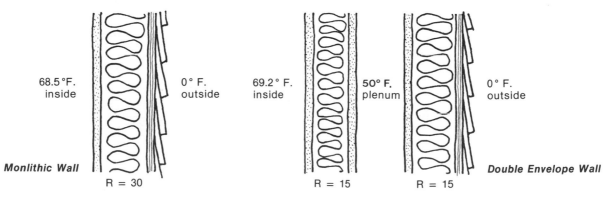

68.5 °F. inside   0 ° F. outside   69.2 °F. inside   50° F. plenum   0 ° F. outside

Monlithic Wall   R = 30   R = 15   R = 15   **Double Envelope Wall**

Comparison of inner wall surface comfort.

directly into a massive rock or block storage maze in the former crawl space and simply superinsulate a monolithic east, west, and north wall? And at the same time, superinsulate a monolithic roof shell and eliminate the attic plenum?

Use a Fan?

"Natural" convection is part of the seduction of the original double-envelope concept. It goes with the arrows and whole wheat, tofu, and a rekindled brand of Yankee independence that exalts self-sufficiency and thrift. Utility companies are to be tolerated up until the machinery building the house is unplugged. From then on the relationship is adversary. Fair enough, perhaps, but what are the real implications of "natural" convection vs. fan-powered circulation?

At seven cents per kwh, a third-horsepower blower can move all the air moved by "natural" convection in an average-sized double-envelope house for about $15 a year. If insulated metal ducts can take the place of the massive attic and north wall plenums, several thousand dollars are saved in initial construction costs. The cubic footage used as plenum area is now available as extra usable living space, and the outbound heat loss from the plenums is eliminated.

Another advantage to the use of mechanical circulation occurs on sunny days, when air temperatures at the top of the sunspace can be over 100° F. At natural convection rates, this hot air will convect itself northward into

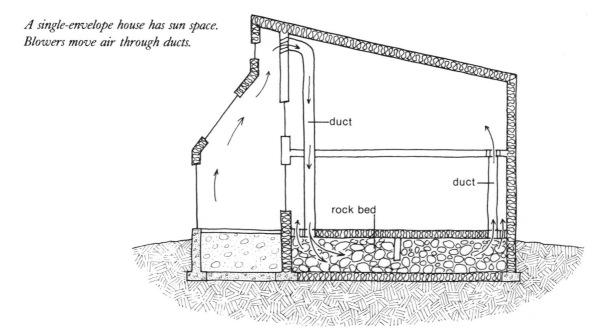

A single-envelope house has sun space. Blowers move air through ducts.

duct

duct

rock bed

the attic plenum at rates up to 100 cfm. Not a bad rate, to be sure, but not good enough if it can be pumped into storage instead of lost to the outdoors while it waits, stacked up against the glass, for natural convection to take it around the loop. Remember that while the sunspace is a huge solar collector, it's also glazed (usually double glazed), and therefore subject to massive heat loss.

R-values for double glazing range between 1.5 and 1.8. (Six inches of fiberglass is rated at R-24.) The higher the temperatures in the sunspace, the greater the ΔT between inside and outside. As the ΔT increases, so does the rate of heat transfer.

Cut Heat Losses

Set to operate at 80–85 ° F., mechanical blowers can reduce sunspace heat loss by limiting those extreme ΔT's. The sunspace-as-collector is, in a way, a paradox. Extremely high temperatures are increasingly useful as heat gain but at the same time self-defeating because of the dramatic rise in heat loss through the sunspace glazing. The energy (electrical) consumed in moving the air quickly to a storage bed is far less than the energy (heat) lost through the sunspace and attic outer envelopes. In the November 1980 issue of *Solar Age Magazine,* Joe Kohler and Dan Lewis offer convincing computer simulations that suggest directly ducted sunspace-to-rockbed systems outperform the convected air envelope system.

Planning Practicalities

While the fluid dynamics of the envelope theory may be subject to numerous explanations, architectural expression has been limited. The modified cape or saltbox configuration is the most commonly used form. It adapts itself conveniently to the southside sunspace and is a culturally familiar package. The north wall of the house is often partially buried into a berm or hillside, a move made especially convenient because of the absence of windows on that back wall (as well as the obvious benefits of earth moderation and zero infiltration rates).

Household usership of the sunspace is the key to the organization of interior spaces. The sunspace typically runs the entire length of the southern (priority) exposure and is often the only source of light, view, and "outside" air. Access to the sunspace is, of course, preferred for all the primary spaces but is unfortunately unavailable to some or all secondary spaces.

This leads to the same dilemma inherent in the design of elevational earth-sheltered houses. The larger the house, the more difficult it is to arrange room after room along the south wall without creating a long hotel corridor connecting the rooms. The double-envelope concept lends itself easily to a two-story building (which gives it some advantage over the typical single-level earth-sheltered plan) so that the building will tend towards a long, narrow two-story form. The double level provides twice the amount of access to the sunspace as the single level and increases the vertical dimension of the sunspace for optimum solar input. The disadvantage is that all the primary rooms open into one "outside" area. Privacy and variety (sunrise, sunset, views, breezes, fragrances) are severely limited. The person in bedroom two who likes to sleep with the windows wide open will either miss the night-cooled sunspace air or be kept awake by grandpa, in bedroom three, who snores like a water buffalo.

The sunspace is often used as a greenhouse as well as a collector and part-time living space. Because it gets cold—into the low 40°s F. at night—the greenhouse-related moisture will invariably condense on the inside of the outer glazing on cold winter mornings. As soon as the sun warms the glass, the condensation will vaporize back into the air. If the weather is cold and grey, the glass may remain clouded with moisture all day. Humidity usually runs around a very comfortable 60 percent because of the plants and earth-coupling in the crawl space below. Condensation on the glass is a symptom of high humidity. It certainly has no aesthetic appeal. Blowing a stream of air across the glass surfaces will help eliminate the problem but will require energy consumption. Detailing the jambs and sills against condensation/water damage should be done with care.

Potential of Sunspace

As an architectural device, the sunspace holds considerable untried potential. The inboard wall can be curved or angular, floor levels can change, balconies and intermediate platforms may protrude and retreat into the house, and roof lines might be juggled according to inspiration. The two-story aspect of the sunspace is a treat to work with because it's the only justifiable use of vertical space available to the energy-conscious de-

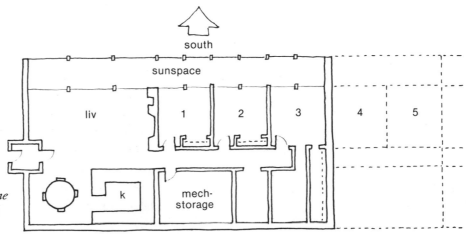

The hotel corridor syndrome inherent to the linear plan development.

The sunspace (at left) of the Bast-Carlsmith house in Hinesburg, Vermont, provides the heat that circulates through the double-envelope structure. Right: As in most double-envelope homes, the sunspace offers a place for growing plants and family living.

signer. Most high-ceilinged rooms are frowned upon as impractical heat traps, but added height in the sunspace increases collector capacity and adds drama and variety to the interior.

The exterior of the double-envelope house is somewhat architecturally confining. The huge expanse of sloping south glazing tends to flatten the elevation to a predictable degree of severity. The north, east, and west walls promote little potential for expression due to minimal or non-existent windows and, in some cases, partial submersion into the ground. The cross section demands a high south wall and, inevitably, a low north wall. Because the building is heliotropic, the sun-funnel form is a fundamental requirement. In a purely sculptural sense, there is nothing abhorrent about a wedge-shaped building, and yet its unidirectional emphasis restricts its aesthetic character to the unfamiliar.

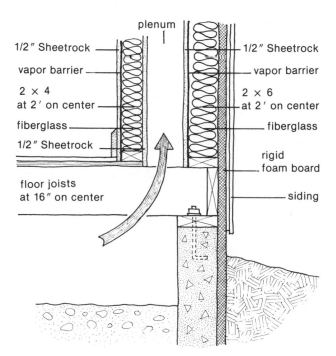

1/2″ Sheetrock
vapor barrier
2 × 4 at 2′ on center
fiberglass
1/2″ Sheetrock
floor joists at 16″ on center

plenum

1/2″ Sheetrock
vapor barrier
2 × 6 at 2′ on center
fiberglass
rigid foam board
siding

A typical double-envelope wall.

How It's Built

The double-envelope house employs the same stick-built (balloon and platform framing) techniques used in 90 percent of the residential work done in this country for the past seventy-five years. Once the foundation is in place the walls are raised and plumbed, followed by the roof and roof waterproofing.

Next, the outside walls and ceiling are insulated and (typically) covered with gypsum wallboard. The inner envelope walls and ceiling are then raised and insulated. The inner walls will have wallboard fastened to the outside face before they are tipped up into place. The space between the two envelopes is too narrow (nine to twelve inches) to allow working room *after* the inner wall is in place.

The inner ceiling is a different matter. The air space between the two roofs is generally wider than the back wall air space and is often deep enough to allow a carpenter to wiggle around and swing a hammer, applying the wallboard to the air-space side of the inner ceiling after the framing is in place. This is tedious work, and could be replaced by a technique similar to the inner back wall sequence, although the wallboard and extra framing member weight would make a horizontal tilt-up operation awkward at best.

Once the inside walls and ceiling are in place, insulation is installed and the vapor barrier applied. The reasons for applying wallboard to both faces of the convective-loop air space are twofold. The wallboard provides a fast and inexpensive method of minimizing turbulence and friction within the air space. It also provides a moderate degree of fire protection to an area whose chimney-like dynamics could be dangerous if a blaze were to begin within the loop. Some double-envelope home designers have added fusible-linked flaps which close off the loop if a fire breaks out. (The venerable William Shurcliff is fond of goading the double-envelope enthusiasts with the question of how to retrieve the trapped, stubborn—or dead—house pet from the inaccessible reaches of the air space.)

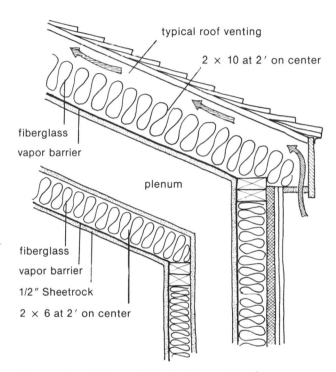

Construction of the roof of a double-envelope house.

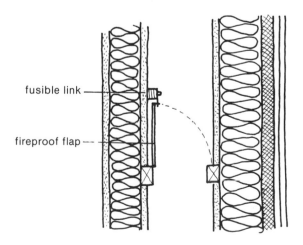

Fusible-linked flaps in the plenum will drop down and halt air circulation in case of a fire.

A tremendous sunroom links the wings of the Minergy House in Massachusetts.

Inside the sunroom of the Minergy House, a hot water panel installed for preheating the domestic hot water supply.

90

The Debate Continues

Double-envelope home designers are an inventive and conscientious lot. The concept is new enough and subtle enough to permit a variety of interpretations to pragmatic evidence, and controversy abounds. The debates continue, healthily, as of this writing, and the most recent design work, such as Hank Huber's, has been eagerly adapted to the evolving pattern of data.

Because double-envelope houses are the most dynamically complex of the new passive building types, the tinkering potential is irresistible. Dozens of these houses are wired up with sensors placed at every conceivable intersection monitoring air flow rate, direction, and temperature, earth temperature, and moisture content. The recorded data are voluminous and the results are freely exchanged among the design community in a spirit of common enterprise. Within a few years, the tempest will have stabilized and the double envelope will have matured into its true form. Meanwhile, we can nourish the process and relish the controversies.

5

SUPERINSULATION

In hard economic terms—comparing capital costs to operational savings—the superinsulated house is the most frugal design available today. Superinsulated houses are conceptually simple, stylistically flexible, and easy to build. They tend to be disarmingly practical. Unlike their earth-sheltered and double-enveloped counterparts, they are woefully lacking in charisma. Unfortunately, they don't interact in mysterious ways with elemental forces. They don't rely on "natural" convection, thermal storage mass, and geothermal tempering. They don't use "free" solar energy. They can be sited facing due north if the site demands. The superinsulated house doesn't do *anything* in the kinetic, complicated way we expect of our radical alternative technologies. Superinsulation is a boring, predictable, prosaic concept that works so simply and so well that it almost takes the adventure out of pioneering.

When we decided ten years ago to design energy-efficient houses, we promised to replace oil and coal with solar, hydro, and wind power. We passionately promoted the wisdom of living symbiotically with nature, of integrating her free, renewable resources into our petroleum-jaded lives. Solar panels, wind turbines, and photovoltaic cells became the high-tech solutions to energy independence. The potential was—and is—exciting . . . almost as exciting as the whirring, clicking, ticking machinery and fractious debates that accompany every new breakthrough. We love gadgetry and we're intrigued by a hint of magic. If something works well, it should be clever.

We're fascinated with animation, even if it's only diagrammatic—as with the red and blue arrows in the dance of the double-envelope loop. How, then, can we expect to get excited about a pudgy, thick-shelled box—a superinsulated house—when so little whirring and ticking goes on? Passive solar design may be dull, but at least we can play with Trombe walls, water storage tanks, and convective air loops. A superinsulated house is inert, the ultimate in passivity, a stuck-in-the-mud ox cart compared to the double-envelope Ferrari, but it gets the job done, and does it well.

High R-Values

As the name suggests, superinsulation is a matter of doing to excess what we've previously done in moderation. R-values are escalated as high as R-40 in the walls and R-60+ in the roof. Thermal resistance is so high that the building is warmed with cast-off heat from lighting fixtures, appliances, and warm-blooded occupants. This "internal gain" is common to all inhabited buildings and may amount to well over 100,000 Btu every twenty-four hours in an average household.

Superinsulation values are chosen to match internal gain so that little or no auxiliary space heating is required. The $3,000 to $5,000 normally spent on a residential central heating system pays for most of the superinsulation treatment. While the cost of heat produced through internal gain is traditionally assigned to budget categories such as "lights," "hot water," and "food," each of

The sunspace in the Brookhaven House is a double-glazed kit greenhouse. It helps heat the house, and provides useful additional space for family activities as well as growing plants.

these forms of energy disperses heat as a by-product of the work being performed.

The old Vermont farmer used a light bulb as an inexpensive portable heater for his chicken incubators. The bulb filament is a slightly sophisticated form of resistance heating, generating far more Btu (heat) than lumens (light). Even *insulated* electric hot water heaters lose 10 to 15 percent of their heat to the surrounding air. Hot water pipes—typically uninsulated—may lose an equal amount, depending upon the location and length of their runs. The people, dogs, and cats around a house produce heat as they metabolize food. The energy potential in a tuna fish sandwich may be 500 calories. Some of that energy will be converted to work (horsepower) and some will be cast off, in the process, as heat (Btu). Obese humans are low-performance consumers, converting energy-rich fuel (food) into fat rather than productive work. Cattle and pigs are purposefully bred to store energy in their bulk, later to be converted, as food, to work.

Heat of Dairy Barn

The modern dairy barn is a good example of the efficacy of internal gain. During the winter, when the cows are kept inside, the ambient temperature must be kept no higher than 54° F. in order to impede the growth of harmful bacteria. With moderately insulated walls and ceiling, the animals produce so much heat that ventilating fans are kept running—even in the coldest weather—to keep the temperature *down* to 54°. And in the milk room, heat radiated from the bulk tank keeps water pipes from freezing—with no auxiliary form of space heating.

In primitive agrarian cultures, sharing the house with the animals was a practical matter—a sharing of B.T. Moos. A "three-dog night" is a reference to three furry bed partners.

Waste of Energy

About 40 percent of the energy consumed in an average underinsulated American home is used for purposes other than space heating. The work performed by this 40 percent is accomplished at far less than perfect efficiency. Electric refrigerator motors, for instance, operate at 50–60 percent efficiency. The balance is lost to friction, hence heat, hence a contribution to internal gain.

While some appliances work at high levels of efficiency (a blender), others are unforgivable energy hogs. The "frost-free" refrigerator uses almost twice the kwh as the manual defrost models. Why? Because the frost-free mode activates an *electric heater* to remove frost-producing vapor! This perversion of purpose will someday top the hit parade of history's most offensive energy omnivores. Most internal gain sources are inevitable by-products of thoughtfully designed but inherently imperfect machinery. Frost-free refrigerators (and "self-cleaning" ovens) are beyond forgiveness.

Old Idea Revived

The old notion that internal gain could account for space-heating requirements within a superinsulated shell found renewed interest at

COST FOR RUNNING ELECTRIC APPLIANCES

ELECTRIC APPLIANCES	AVERAGE ANNUAL COST
14 cu. ft. manual defrost refrigerator	$ 25–40.00
17 cu. ft. frost-free refrigerator w/top freezer	$ 50–95.00
Dishwasher, 8 loads per week	$ 65–90.00
30″ range	$ 34–42.00
Clothes dryer, 8 loads per week	$ 44–48.00
Water heater, 65 gals per day @ 140° F.	$300–370.00
Electric furnace, 50,000 BTU/hr @ 2080 heating load hours per year	$1170.00

the University of Illinois in the early seventies. A group of architects and engineers, including U. S. Harris, R. A. Jones, S. Konzo, and W. L. Shick, formed an association with the Small Homes Council and formulated the groundwork for superinsulated design. Adding the results of computer analysis to their theoretical work, they published plans for the Lo-Cal house in 1976. By the end of the decade, dozens of superinsulated houses had been built, inspired by the Lo-Cal research team and their Canadian counterpart, the Saskatchewan Research Council. The Saskatchewan group built a prototype in 1977 and proved the concept could perform well in harsh (10,-800 degree-days) locations. Their house in Regina, 1,000 miles north of Chicago, included an active, drain-down, flat plate solar collector, which was to have supplemented internal gain for space heating. It was found that the active collector was not required. At minus 30 ° F., the heat loss per hour is a meager 12,600 Btu, an extraordinary figure for a 2,000-square-foot house.

A Low-Tech Affair: Vapor Barriers and Low Infiltration

Building a superinsulated house is a low-tech affair. The only areas of concern have to do with vapor barriers and low infiltration rates.

Vapor barriers are essential in any high-performance house, doubly so where large quantities of fiberglass or cellulose fiber insu-lation are used. Both materials will absorb and hold water. In cold weather, moisture-laden, warm interior air works its way through any permeable material as it seeks to equalize density with the dry, cold outside air. Without a vapor barrier, a typical wall section composed of gypsum wallboard, fiberglass insulation, and exterior sheathing will be subjected to enormous quantities of vapor.

The negative effects are considerable. First, insulation loses its insulating value when it becomes wet. Water is a good conductor, thus a poor resister of thermal transfer. A cotton shirt or a down-filled jacket keeps us warm as long as it's dry. Soaking wet, neither garment is effective. Building insulation is subject to the same principle. Vapor can accumulate to such a degree in a fiberglass or cellulose-filled shell that the R-factor will be reduced to a fraction of its original value.

Saturation culminates as vapor reaches the *dew point,* that critical zone in which vapor turns to liquid as it proceeds from the *warm* interior towards the *cold* exterior surface of a wall or roof. If the sequence of vapor-to-water is repeated often enough, as it will in cold weather, water will saturate the insulation intolerably. In extremely cold weather, water will freeze in the insulation and the buildup of ice will further restrict the passage of vapor from warm to cold.

The closed-cell insulations—urethane, urea-formaldehyde, and polystyrene—absorb very little vapor and are, in fact, fair vapor barriers themselves. Urethane is often manufactured with metallized foil bonded to each surface,

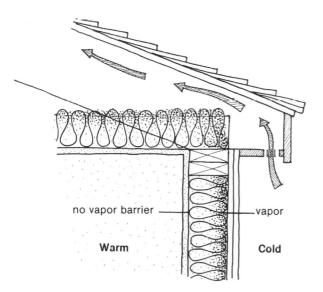

This wall lacks a vapor barrier. Vapor penetrates insulation, turns to liquid when it reaches cold air. Result: value of insulation drops, water rots wood.

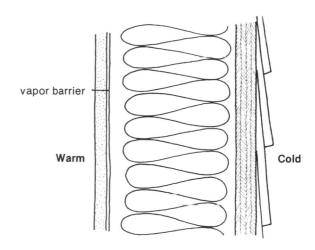

Vapor barrier must be placed as close as possible to the heated area

reducing permeability to near zero. The joints between the panels then become the weak link in vapor control, but these can be sealed with metallized tape carefully applied. In practice, the use of rigid foam insulation is usually preempted by the more economical fiberglass or cellulose fiber, but when it is used, a contiguous, additional barrier is still advisable.

Polyethylene Film

Polyethylene film is the material most commonly used for vapor control. The six-mil thickness is recommended for its strength against punctures or tearing during installation. Care in application is of paramount importance because even the smallest hole will allow moisture to begin its route towards the dew point. The barrier is applied to walls, ceiling, and floor only after all framing, insulation, rough plumbing, and wiring have been completed. Stapled into place and generously overlapped at the joints, a properly installed vapor barrier will be installed as close to the heated space as possible. In most houses, this means the barrier is directly behind the wallboard. It should be gently stretched, uninterrupted, over windows, doors, and electrical boxes and then, wherever necessary, cut away and taped at the edges to form a monolithic seal.

If the vapor barrier works perfectly, the house will perform at its optimum. If the vapor barrier fails to some degree (and we assume that it won't be perfect), we can offset the potential damage with ventilation.

The roof should be ventilated in any heated building that uses fiberglass or similar vapor-susceptible insulation. A vented roof system allows moisture to escape to the outside. An unvented roof forms a trap.

Traditionally, walls have been unvented and have, so far, escaped the vapor-trap reputation associated with unvented roofs. Walls are typically insulated to between a half and two-thirds the R-value of the roof (more heat—and vapor—rises than moves laterally) and therefore provide less of a "sponge." However, the advent of superinsulated means the advent of supersponge. Detailing that allows the walls to "breathe" may need to be more rigorously developed.

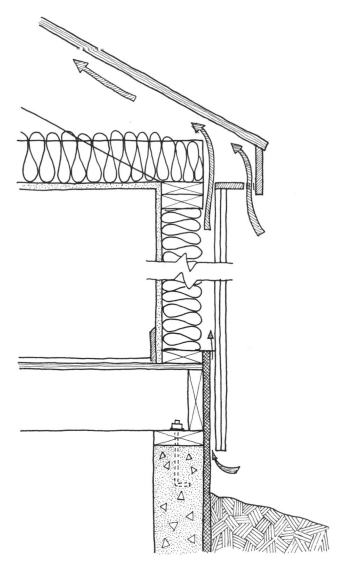

Wall venting into roof system.

Breathing can be assured by means of an outboard airspace directly routed to outside air, as in a typical roof system, or a sheathing material with a high perm-rate. Board-and-batten siding, for instance, breathes very well. Plywood, however, because of the glue between the laminates, is relatively impervious. Blistered paint on wood clapboards is invariably a sign that the wall is suffocating. Vapor pushes out through the clapboards and bubbles the paint. The most severe blistering is usually found outside bathrooms and kitchens where vapor levels are highest.

Wood May Rot

If the wall or roof is not vented and the vapor barrier does not function properly, moisture in the insulation not only minimizes the R-value, as discussed above, it also sets up a perfect environment for rot. Studs, sills, ceiling joists, and rafters can deteriorate in a few years if they're trapped in an ever-moist environment.

Vapor barriers are supposed to eliminate the need for venting, and in theory, they do. But the discrepancy between theory and practice may be accompanied by dire consequences in a superinsulated house. My advice is to vent it, all the way around, and be safe.

Air Changes

When the wind blows, a house shivers. Not only is the wind pulling heat away from the surfaces of the building (wind chill) but the

A modified Trombe wall plus paving brick such as used to build this arch combine to provide thermal mass for heat storage in the Brookhaven House, funded by the Department of Energy through Brookhaven National Laboratory, Upton, N.Y.

increased wind pressure forces air into the building through cracks that even water wouldn't penetrate. This wind-borne assault is the extreme form of infiltration and is measured in terms of air changes (per entire house) per hour. In a fifteen-mile-an-hour wind, an old colonial farmhouse might undergo five to ten complete air changes (transfusions) an hour. Replacing stove-warmed air with frigid north wind cost our grandfathers more than twenty cords a winter. Infiltration—wind blowing through the house—has traditionally been one of the major areas of consideration in heat loss calculations.

High performance houses have conquered the old problems of infiltration with earth coverings, superinsulation, multiple shells, and modern materials. Weatherstripping and caulking around windows and doors have eliminated most of the remaining points of access.

Providing Fresh Air

But the old problems have been replaced with a new one, especially in the supertight, superinsulated houses. Infiltration is a source of heat loss; it is also a source of fresh air. In an effort to perfect performance, the superinsulated design minimizes infiltration rates to a dangerously low level. It's generally agreed that an air change rate as low as .5 per hour (half the air in the house is replaced by outside air every hour) is the minimum that is acceptable. An air change rate of .2 an hour is considered unacceptable. Who wants to live in a house with bacon breath? Not only is stale air (housetosis) unpleasant, it also becomes a health hazard when it's left to incubate the petri dish of household bacteria found in all but the most compulsively scrubbed split-level condo.

101

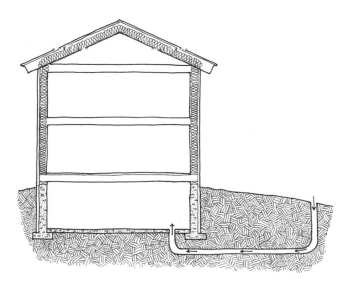

Earth pipe will bring in earth-tempered air.

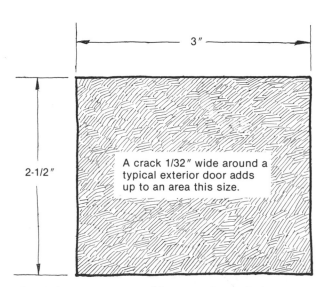

A crack 1/32″ wide around a typical exterior door adds up to an area this size.

Cracks in house may provide too much ventilation

Low infiltration rates mean lower heat loss, but is this kind of economy worthwhile? Even a year's supply of fuel oil doesn't cost as much as a week in the hospital. Fresh air is as much a part of an aesthetically pleasing home as the walls themselves. In the coming generation of energy-efficient houses, we will see increasing numbers of air-to-air heat exchangers automatically providing fresh outside air. Superinsulated and earth-sheltered houses are most apt to require mechanical air supply control. (The double-envelope configuration relies on the sunspace for adequate infiltration, accessible second-hand through an operable window in the inner shell.) The exchanger collects the heat being exhausted, trading warm, stale air for cold, fresh outside air. Not a big bargain, but it does raise a question of costs. Is the purchase price (up to $1,000) and operational cost ($10 per year) of the equipment worth the Btu saved?

It may be more economical to allow for an optimum amount of fresh air to infiltrate the house naturally. This can be achieved with carefully calculated, adjustable air slots protected from the wind. *Earth pipes* are another popular and more expensive solution. Outside air is brought a long distance through a culvert or pipe buried below the frost line. The tube's contact with the earth moderates the temperature of the air passing through. A nice idea, but the payoff is cost-ineffective compared to alternatives.

The simplest solution is to install a few doors or windows without weatherstripping.

Or hire a sloppy carpenter to install the doors *with* weatherstripping. The problem with that is the inability to adjust according to winds, number of occupants, showers, and cooking schedules.

Wood stoves, furnaces, or fireplaces in any house should have an outside source of make-up air for use in combustion. A four-inch round duct is the smallest that is adequate for most applications. It should be protected on the outside end with an adjustable valve.

Framing Methods

Superinsulated houses typically use standard balloon or platform framing systems with an added inner stud wall. The inner wall is not bearing a load, and is separated by an inch or more from the outer wall. The massive amount of insulation between the two walls is thick enough to make the twin walls (2 x 4s at sixteen or twenty-four inches on center) more economical than a single 2 x 10 stud system in initial building costs as well as energy performance. A traditional stud wall loses a fair portion of its R-value to heat transfer through the studs. Three and one half inches of fiberglass batt insulation are rated at R-13 in a wall. The fiberglass batts are 14½ inches wide, fitted between studs 16 inches on center. The stud itself has an R-value of about four. This means that 1½ inches out of every 16 inches in a standard stud wall (about 10 percent including sills and

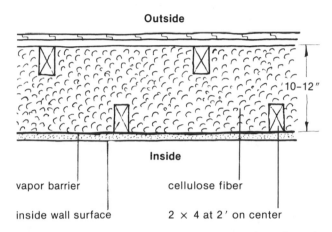

A typical superinsulated wall doesn't transfer heat through studs.

A heavy coat of insulation wraps the Brookhaven House at levels of R-27 for the walls and R-38 for the roof. Stud walls are made up of 2 x 6 studs every 24 inches with full-depth batt insulation. Two inches of rigid foam insulation cover the foundation walls, from the footings to the top of the foundation.

plates and headers) rate an R-value of four rather than the thirteen (plus or minus) calculated for the insulation. Furthermore, the batts are bound to fit less than perfectly between the studs, allowing for possible infiltration twice every sixteen inches as well as at the top and bottom.

The cavity between inside and outside walls is filled with either multiple layers of batts or poured-in cellulose fiber. The batts offer a lower R-value and must be carefully fitted. The less expensive cellulose fiber conforms extremely well to whatever container it's poured into. If the cellulose fiber is properly installed—gently packed down—settling is not a problem.

Roof Systems

The superinsulated roof system is very simple if the insulation can be piled up across the ceiling joists. Providing usable space above the insulation is expensive. It requires additional structurally redundant framing—like the walls—and decking due to the unusual depth of the insulation. An R-60 roof insulation application will be at least eighteen inches deep, vent space included. Superinsulated houses don't easily lend themselves to attic space. They strongly suggest a plan that avoids double framing. This means flat roofs, low-pitched shed roofs, or gabled roofs with

104

inaccessible attics. Venting above the insulation is essential.

Can It Be Beautiful?

What about architecture? Can the superinsulated house be made beautiful? Can a conceptually inert design aspire to aesthetic expression? Of course it can. Despite the overwhelming evidence to the contrary, the superinsulated concept imposes few restraints on the designer. I have no idea why the prototypes are so uniformly ugly, but there are no inherent limitations other than fenestration (the treatment of windows) and a reasonably low ratio of volume to surface area. There are no large glazed openings in the classic superinsulated designs, because although windows admit a certain amount of solar gain, they are also a primary source of heat loss. Since the superinsulated scheme depends upon internal gain for space heating, windows become thermal liabilities. Occupants aside, the ultimate superinsulated solution would have *no* windows, a totally unacceptable solution from a humanistic point of view. Most forms of architecture rely on the treatment of windows as a primary organizing device. In the superinsulated house, the window may still be central to form and space, but it will occur as a specific incident rather than a series, as punctuation rather than syntax.

A mitten warms the hand better than a glove because a mitten has a lower ratio of

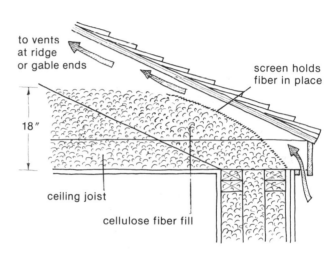

A wall and cap such as this add up to R-60 roof insulation.

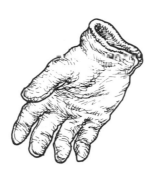

Both enclose the same volume, but glove has twice the surface area of mitten. Architects call it volume-to-surface area, and watch it carefully when considering heating problems.

enclosed volume to surface area. The finite amount of heat being generated by the hand is dispersed through fewer square inches of material. A globe is the optimum volume-to-surface shape, but a difficult configuration to build and live in (Buckminster Fuller notwithstanding). A cube or a shoebox shape is the most practical, habitable form that recognizes the volume-to-surface principle. A building plan designed in the shape of a capital H is doomed long before it begins to count calories. Any of the building types we have discussed will perform best if the volume-to-surface ratio is kept within reason. The double-envelope and the elevational earth-sheltered designs are partially exempt insofar as they rely upon a long south wall for passive solar collection. But the superinsulated concept relies upon the potential of internal gain within an optimally insulated shell and needs no long axis towards the sun. Its shape is determined by internal functions and whatever aesthetics the designer chooses to work with, but the chunkier, monolithic forms will perform best.

The superinsulated house will never be identified with the trendy architectural iconography emblematic of "solar" or "high tech." It is an inherently conservative concept with predictably limited forms of expression. With skill, it can be handsomely rendered. Its architectural potential is still virtually untapped. The principles are solid. Superinsulation will surely become one of the few universal building blocks in the future of energy-efficient design.

PROCESS

6

The mysteries of the creative gesture have been long belabored and little understood—perhaps least understood by artists themselves. Knowing more about how we conduct our everyday lives may bring self-realization, but knowing how and why the lightning bolt of creativity strikes is an enigma. The working artist is not much concerned with these questions as long as inspiration flows. The "blocked" artist, on the other hand, will agonize over the missing muse and try desperately to re-create the conditions under which his or her last great creative moment occurred—usually to no avail. Like the frustration of trying to remember a name or trying to get to sleep, trying to be creative (like trying to be spontaneous or trying to *not* try) is an exercise in futility. The Big Idea comes when it's good and ready, in the shower or half way across a shopping center parking lot, but it won't be rushed and it will never repeat itself.

I'm inclined to dismiss discussions on "creativity" as solipsistic. Artists need to know why they create art about as much as birds need to know about ornithology. I don't mean to be arrogant or precious about sharing admission to the inner workings of the creative mind. I've simply never found a ticket to that particular show, and I'm not convinced anyone else has, either.

Value of Process

On the other hand, I'm enthusiastic about the value of a firm understanding of *process.*

Process can be learned, refined, and perfected. Process is the art of work, whether contemplative or manual, and it's a form of art in itself. When combined with profound insight and clarity of expression, process, the art of work, can lead to the work of art. The work of art—the artifact itself—is traditionally identified as the valuable commodity, while the means to the end is rarely found on a pedestal. In the performing arts—dance, theatre, and music—process and product are identical. The pyramids will last forever, but a brilliant leap by Nureyev is recorded in memory alone the moment it's been accomplished, process and art inseparable.

Architecture is peculiar among the arts in its need to perform a number of complex practical functions as well as to gratify the eye and soul. The architect's mundane restraints are structural, mechanical, and economic. Form, space, and light must serve—and be served—by practical demands unfamiliar to other artists. Furthermore, the architect works for a client whose tastes and pocketbook may substantially alter the development of an otherwise perfect (abstract) architectural solution.

Because of its complicated nature, architecture *must* rely heavily upon a rational process of design. The architect's "feelings" for a solution constantly undergo a rigorous cross-examination from all corners. Will the windows allow enough light into the building? Do the walls provide adequate lateral bracing? Can a column be omitted? The answers to these questions have aesthetic as well as prac-

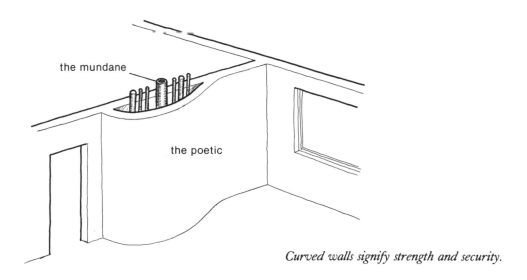

the mundane

the poetic

Curved walls signify strength and security.

tical implications. The architect is committed to a process of constant compromise between the prosaic and the poetic. Although it may be no more than wishful thinking, architects like to believe that the ideal solution forges a symbiotic union between form and function, each side enhancing the other. In reality, or what passes for reality, a gracefully curving, pristine wall may be a convenient wrapper around a twisted jumble of pipe and wires. Fair enough, perhaps. Accommodation has been made to both the eye and the thermostat. Other examples: a structural column doubles as an air duct, a tiled floor acts as a heat storage mass, and a skylight provides a view as well as solar radiation. The decisions leading to these kinds of solutions are a result of deliberate, disciplined thinking about how things go together, a process of simplifying the complicated, synthesizing the parts into a whole.

Prejudice

Prejudice has a distasteful connotation in our society. It's despicable in its traditional forms, but in the realm of design it has an important role to play. Despite my dedication to the idea of an open-minded, tool-oriented approach to design, I'm quick to confess that we're not likely to proceed *tabla rasa* into any experience, house building especially. We're hardly innocent of likes and dislikes in colors, textures, materials, and forms. We've stored away a lifetime of impressions labeled "ummhh" and "ugh," and we deny ourselves unnecessarily if we ignore our intrinsic aesthetic sensibilities.

The trick, of course, is to recognize the difference between sensibilities and habit. Sensibilities are part of our intuition and can embrace innumerable variations. Habits tend to be applied behavior, unaccommodating and rigid. Habits we can break. We can learn to enjoy hardwood floors instead of linoleum. Our prejudices, our sensibilities, however, are made of sterner stuff, and we may as well face them as an inevitable component of our design vocabulary. Aesthetic prejudice is a springboard. It always needs testing but it never fails. It can't work alone, but it's vital and healthy and we're never without it, so we may as well cultivate it as an indelible resource we can trust and enjoy.

Inspiration or . . .

When I think of the process involved in designing a building, I try to ignore inspiration (it will come . . . I hope) and depend, instead, on a) tools, b) systems, c) the aforementioned prejudice, and d) a stubborn insistance on evolution.

Process feeds on evolving ideas. "If this part works here and those pieces want to be over there, then *that* begins to suggest . . . ." Evolving ideas are merciless when they reveal the previous day's work to be an extinct species, but once evolved, a new idea deserves to be tossed around as rigorously as its predecessor, until it fits its function or begins to suggest . . . an even further refinement. It takes courage and an active interior dialogue to nurture an idea to maturity, but the rewards are worth the effort. To be willing to ask . . . "What if?" over and over again is an invitation to the opening door of creative intervention. When inspiration pays its visit, it invariably follows the question, "What if?"

What if we ask the questions without being aware of existing answers? If we're clever, we may reinvent the wheel. More likely, we'll come up with a square hub cap. It may be amusing to ponder the potential of a windmill, but a little research shows that a given windspeed will produce X number of kilowatts. The cost of the windmill translates into a cost per kilowatt of electricity produced. How does that compare to the utility company rates? What if the utility's rates go up? What if the windmill needs a new widget? What if, what if? Information is crucial to these answers. Information is the fundamental tool in the designer's workshop. Without the tools, the process is severely hampered. False premises beget false solutions. Information adds heft to the tool chest, adds flexibility to the designer's range of options, adds a sharp cutting edge to the inquiry.

Patterns Emerge

As one becomes familiar with the tools of the trade, patterns of use begin to emerge. Just as the hammer is meant to hit the nail, the Btu potential in a gallon of fuel oil is keyed to formulas for heat loss calculations. Sets of tools (information) begin to emerge and add up, the way certain notes in the scale combine to make a particular chord. Systems are made up with these sets of information working together around a central idea. Earth-sheltered architecture, for example, begins with the notion that earth is a benign climate into which a building can be advanta-

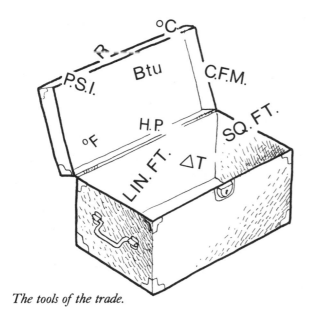

The tools of the trade.

geously situated. The idea seems valid. The next step is to test the idea with the tools of technology, the tools of engineering. Leaks, cave-ins, humidity, and heat loss can all be anticipated with the appropriate information. Once quantified, the information is tested and refined and soon evolves into a system.

Replace Systems with Tools

But what begins as a practical system of earth-sheltered technology often becomes a panacea, rife with quasi-philosophical claims and susceptible to the myopia of all monocausal approaches to life.

The earth-sheltered system is a good system. The double-envelope system is a good system, and the superinsulated system is a good system as well. Each depends upon a system's approach, that is, an approach that develops a central idea to self-justification. The problem is that these systems have isolated themselves from the potential advantages of their companions.

The state of the art of high-performance house design is at the moment hampered by the system's approach. I'm convinced we must return to a less-structured approach in which ideas, guided by information, evolve into solutions free of systematic expectations. The systems currently available to us should be reduced to tools. We need to dissect systems regularly in order to avoid the inevitably misleading dogma that comes with a preconceived ideological package.

If you're determined to build a double-envelope house, you must be prepared to accept the attendant restraints whether they appeal to you or not. If, on the other hand, you want to build a house that responds to your specific site, lifestyle, aesthetics, pocketbook, and prejudices, it's unlikely you'll find it in an over-the-counter package. "Be particular, millions are. . . ." was an ironically popular jingle advertising Pall Mall cigarettes a while back. The degree of particularity suggested by that silly campaign is of course self-contradictory, but in a limited way, the systems approach is similarly indiscriminate. If we are, indeed, to sit down and design the best damn house we can come up with, we're talking about using every tool available to us in the most energetic way we know how. Earth-sheltered, double-enveloped, superinsulated, grass hut, wattle and daub, thatched-roofed, igloo—they all show us something that we may be able to use, but none has all the answers by itself. Look around the corner and behold. Answers are everywhere.

SUPERHOUSE

7

A Tuesday morning in March: the phone rang for the fourth time since we got to the office and it was not yet nine o'clock.

Two contractors called about jobs in progress. Could bead board be substituted for Styrofoam? No. Was a piece of steel on the framing plan meant to sit on a steel column or a stud wall? Steel column, thank you. A call from a client. When would she have prints of the interior elevations? Next week— if the phone will stop ringing.

The fourth call was from a stranger, a woman from Connecticut. Her name was Eleanor Guthrie. She and her husband, Walter, were thinking about building a retirement home on a piece of land they owned in New Hampshire, not far from me. They were interested in the new technologies. They wanted to build a house that would require as little energy input as possible. They had been reading about superinsulated houses and underground houses and double-envelope houses. She knew she wanted to take advantage of whatever was available, but she was confused as to which way to go.

Familiar Introduction

Her introduction sounded so very familiar. Most of my clients over the past six or seven years have been adamant about the idea of energy independence. "Spend a little more now and save a lot more later. Let's build it right. It's the last (only) house we'll ever build. What can we do to make the building energy-efficient as well as beautiful? Our kids have grown up, our old house is too big and drafty, and we're sick of paying the utility companies and painting contractors. We want to live a little closer to nature and enjoy the good life without feeling guilty about using more than our share of the planet's natural resources. We want to participate in the solution to the energy problem. We're willing to sacrifice a little—money, comfort, lifestyle— in order to be part of a just conclusion to this energy mess. We're willing to be flexible, within limits. We want to be open-minded about any sensible concepts you might want to try on us . . . and we'll need a spot for our antique corner hutch and a way for our tabby cat to let himself in and out without waking us. . . ."

Familiar Conversation

As we spoke, my partner, Geoffrey, looked over from his drafting board. He had listened to my end of these conversations many times before. He heard me downshift to a patient crawl as I fielded the questions so new to Eleanor, so predictable to me. Yes, we were aware of phase-change salts. Active solar systems in New Hampshire? We would talk about the pros and cons when the time comes. Her questions were intelligent, her interest seemed genuine, and she sounded as if she and her husband would be reasonable people to work with.

I closed the conversation with some questions of my own. It was important for me to gauge her degree of architectural ambition, as well as her commitment to energy-related issues. I mentioned architects and buildings I admired and asked for her reactions. She was enthusiastic about some of the same things that excited me. It sounded as if we were aesthetically compatible. She and Walter would interview us (and we would interview them) at our office later in the week.

As I said goodbye, Geoffrey and I mimicked one another—our habitual parody of the hopeful, self-defensive gesture, eyes to heaven, cynical grin, palms upturned . . . maybe this could be a real one . . . maybe.

Compatibility?

Good architecture requires a close collaboration between a good architect and a good client. The design process is too precious a commodity to share with the wrong consumer. It takes experience and intuition to spot, in a few hours, the critical characteristics that may support or sabotage a job. In some cases there are simply no grounds for compatibility. I don't have the skills or interest, for instance, to design a split-level raised ranch. It would be a waste of everyone's time. Nor can I guarantee my clients exactly what I'll come up with except that it will probably involve the vocabulary I've evolved through my previous work. A certain degree of trust is required of the client, and a leap of faith made when the contract is signed. This excites some (the lucky ones) and terrifies others.

A client's trust in his architect facilitates the long, creative dialogue required of a successful project. On the other hand, the architect's accurate assessment of the client's in-

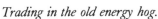

Trading in the old energy hog.

tentions is also crucial, and can mean the difference between a mutually productive job and a year's worth of missed connections. There's no guaranteed way of predicting future compatibilities, but a rigorous courtship provides plenty of clues as to the future of the marriage. Trouble is, the courtship is usually a quickie, followed by contractual vows 'til cost-overruns-do-us-part.

The Guthries showed up promptly at two Friday afternoon in a brown BMW sedan. A clue? Was this choice of Bavarian exactitude emblematic of a frame of mind? An immaculately maintained automobile would probably mean a finicky obsession with details later on. Any dents or scratches? I *hoped* so! Or would they expect their house to perform with the same sort of mechanical precision they expected of their German juggernaut?

Good Beginning

We met in the driveway and exchanged pleasantries. So far, terrific; warm, open, mid-fifties, full of life and enthusiasm. Walter commented on my just-pruned apple trees. He loves gardening. He spotted a mouse's tracks in the snow, zigzagging towards the warmth of the office, and laughed about the universal problem of cold feet. "All of us shiver sometimes; it must be good for us." He stopped at the door to pat the dog. "Good pup." His hands knew about animals. As we went in I looked back at the BMW. There was a bag of dirty laundry tipped over in the rear seat and some paper cups on the dashboard. On the back bumper was a faded sticker. It said "Question Authority." I decided I was prepared to get along very well with the Guthries.

We sat around the drafting table and talked about their lives: married for thirty years, three children, two still in college, and two grandchildren. Eleanor had recently given up teaching sixth grade full time, but was considering part-time substitute work once they were settled in their new home. She had recently completed a course on solar design at her local community college. My sense was that the old warning about a little knowledge being a dangerous thing did not apply here. As she spoke, her references to "appropriate" technology were interrogative and wry rather than declarative. A healthy skeptic with a good sense of humor, she would surely keep us earthbound, but her curiosity and enthusiasm for demonstrable truth would give us all the freedom we were likely to ask for.

Walter, a retired forestry products marketing consultant, nodded in silent agreement to much of what she said. They were a team, I felt, and she was confident enough to speak for both of them. Their attitudes and preferences spilled out in the ensuing chatter, and their ideas about what a house should be began to take shape.

They both love to cook. They kid each other; he's a clean-up-as-you-go type, she throws pots and pans with abandon. An effi-

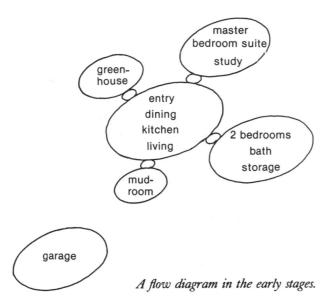

A flow diagram in the early stages.

cient, professionally equipped kitchen is at the top of their list. It should be accessible to the living-dining area. They entertain often and want the cook of the evening to be able to participate in conversation rather than be isolated in a closed-off room.

Walter brings up his ideas of a working greenhouse. He's got two green thumbs—he's all thumbs, Eleanor jokes. We discuss the possibility of a greenhouse connected to the kitchen. He warms up to the topic and is charmingly animated as we imagine the taste of fresh basil in February. Oregano, thyme, and parsley right outside the kitchen door! How about room enough in the greenhouse for a small table and chairs for lunches on sunny winter days? Southern and eastern exposure would provide morning light to the kitchen and maximum input into the growing area. Summer shading, ventilation, and glazing options tumble into the discussion. I promise to return to the particulars once we've refined our thoughts about the house as a whole. Marc Rosenbaum, a solar engineer, will join our venture as energy consultant as soon as our project takes a shape he can respond to.

Project Program

Within a few hours we've developed a program for the project including a few extravagances that we agree to include only if the budget allows. There's rarely enough capital to build everything, so I suggest a "sequence of surrender." The last item to be compromised will be the kitchen. The two most stubbornly held principles will be high insulation values and aesthetic integrity. I take this as a good omen.

The Guthries expect to be by themselves much of the time, with frequent visits by out-of-town friends, children, and grandchildren. We discuss the advantages of a house zoned for such visits, expanding and contracting as occupancy decrees. The visitors' rooms would remain unheated when not in use and would be acoustically isolated from the others.

The plan would be divided into three connected but distinct realms: family/entry—kitchen, dining room, living room, greenhouse—flanked by the private wing—master bedroom, dressing room, bath and study—and guest wing—two bedrooms, bathroom, and storage. The family zone acts as a buffer between the two adjoining wings.

The Guthries agree to round up all their stray thoughts about closet space, washer-dryer location, room sizes, outbuildings (for car storage, tools, firewood, garden tiller, etc.) and type up a formal program.

The Budget

Foremost, but last in the discussion, we talk about budget, delaying the unpleasant topic of financial restraint until the last moment. They hope to keep the cost of the building down to $110,000. I know by experience they will need at least 2,000 square feet for the house they want. At $50 a square foot we would be in under the wire. Site work

(driveway and final grading, seeding and plantings) and utilities (water, sewage disposal, telephone, and electric) would cost an additional $15,000. Outbuildings could cost between $15 and $20 per square foot and they would probably need at least 700 square feet ($10,500 to $14,000 worth) of that kind of rough space. Architects' and engineers' fees would total close to 12 percent of the cost of the building(s).

PAYING THE PIPERS

House	$110,000.00
Site work	10,000.00
Outbuildings	14,000.00
Fees	11,000.00
	$145,000.00

Their land had been paid for long ago, but they expected to invest a fair amount of "sweat equity" in whatever parts of the project they could in hopes of keeping down costs. All totaled, they might spend around $150,000, twice the price of many existing tract houses with equivalent amounts of space. Was it worth it? If the funds weren't available, no. But here, they were.

Morals and Money

We talked about the use of money, self-indulgence, social responsibility, and eco-elitism. Was it morally justifiable to spend an extra $10,000 on an energy-conserving feature that would never justify its initial cost in operational savings? The argument has interesting overtones in its appeal to long-term self-sufficiency. If we spend our cash today on the critical tools of survival, we may avoid a future in which cash has no value, or resources are sufficiently scarce so that they will no longer be traded in traditional markets. Petroleum-based products, for instance—plastics, fibers, medicines, and fertilizers—will claim an increasingly high proportion of the dwindling petroleum reserves. It can be argued that the use of #2 fuel oil will someday become an indefensible waste of a nonrenewable resource. Solar-powered space heating, on the other hand, operates on free fuel.

Like the armchair quarterback and the Mercedes-Marxist, the eco-elitist is locked into an ethical bind, applying privileged standards to problems that others must solve without the benefits of hindsight, trust funds, or social leverage. Drawing the line between elitism and intelligent planning can be difficult.

With the Guthries it was, luckily, not so ambiguous a task. Walter proposed we limit ourselves to a maximum fifteen-year payback on any parts of the project that are designed to save energy costs. We agreed upon a pessimistic scenario as to the rising price of oil, anticipating a 15 percent per year average increase over the next fifteen years.

The other category of "extra" expense—comparing our project with the tract house—has to do with aesthetics. How do we put a limit on the value of a beautiful tile floor or a graceful arch? Eleanor insists she won't live with snap-in-window mullions and plastic countertops. We add clamshell moulding, acoustic tile, and machine "wrought" iron cabinet hardware to the list of contemporary rudeness and reassure one another that we're not that type. It soon emerges that the only constraint on aesthetic expression would be monetary. We would build as thoughtful and integrated a piece of architecture as the budget would allow. If we had to trade some delight for some square footage, the house might end up delightfully smaller.

A Feeling of Trust

What begins to emerge during our interview is a growing feeling of trust among us. We subtly test one another on issues of taste and practical priorities, and find a comfortable resonance in our dialogue.

During an aside into the implications of energy-intensive construction techniques, I see a further dimension of Eleanor's intellectual tenacity. Walter began the discussion by mentioning urethane foam insulation. I picked up on the topic by comparing R-values and costs of the urethane vs. polystyrene. Both are high-tech, energy-intensive products, but counting Btu saved per dollar spent, the urethane is no bargain. Eleanor jumped ahead into the big picture. How much energy are we using to save energy? The energy consumed during the building and decommissioning of an atomic power plant makes that miraculous splitting of an atom not so efficient a

Building materials can be produced with a minimum expenditure of energy.

process after all. The high-tech glazing, plastics, mastics, and metals that make up a solar collector don't grow on trees. The iron ore mined in the West, refined and shipped east, smelted and shaped and shipped back to Colorado in the form of reinforcing bars to be used in the walls of an "energy-efficient" earth-sheltered home have already consumed a generous portion of the energy the home is designed to save. Eleanor points out the paradox of the high-tech, energy-intensive materials and building techniques applied to solve a problem that they themselves help create.

Conversely, the native hemlock roof timbers we use in our New England earth-sheltered houses provide a good example of low-energy production costs. The trees are often harvested within ten miles of the mill, sawed through once (no planing) and shipped green (air-dried on the truck) to the building site, often within the same short distance to the mill.

The equivalent roof system in reinforced concrete will have consumed many times the energy in production and shipping before it becomes a part of its "energy-efficient" destination.

Of course energy costs money, and the ultimate choice of one material over another will usually be made on the basis of economics. High-tech, energy-intensive materials typically cost more, and therefore price themselves out of the competition. But in many cases the costs of the low-tech and high-tech products may be similar, and the issue of how much energy the stuff has already consumed becomes part of making a responsible choice.

Agreed

By four o'clock that afternoon we decide to get married.

It's clear to us that the Guthries are looking for the kind of house we can provide.

123

They, in turn, are excited about the prospect of having us design a house that works for them in the ways they feel are important. We talk about our contract, about seeing the site. We walk the Guthries out to their car, enthusiastic about the prospects of putting together a house with all the essential ingredients. The BMW disappears down the driveway, and Geoffrey and I return to the office, visions of a new kind of house crowding our brains.

By the middle of the next week the program arrived along with a signed copy of the contract and a check for a retainer fee. The program was essentially the same as we discussed the previous week with some further dimensions added.

Guthrie House Program by Walter and Eleanor Guthrie

General: *The house will use all feasible, practical, and proven methods of conserving and utilizing as much free (solar and geothermal) energy as possible. It will attempt to be self-sufficient and maintenance-free. It will unify aesthetics and technology to the compromise of neither. It will be arranged so that it gracefully subdivides into three realms when required, but will operate optimally with the guest wing closed off. It will be integrated into the site rather than impose upon it. It will recognize the natural elements around it and foster intimate access to the beauty of sunrise, sunset, wind, rain, snow,* *moon, and stars. It will be built to last and, we insist, it will be built within our budget.*

Specific: *A single entrance will provide both a welcome to the guest as well as an opportunity for family member decontamination—stowing away muddy boots, rain wear, snowshoes, etc. This would probably be best achieved via a mud room closed off from the entry hall by a doorway just inside the front door. Would this be a good place for the washer-dryer? Utility sink? Will we need a guest closet in the entry?*

The entry will provide direct access to as many parts of the house as possible; there's no reason for children and grandchildren traffic through our activities each time they come and go. We also feel there should be some sense of transition between the point of entry and the living room.

Large kitchen
The kitchen must be large enough for two cooks. We need a range top, two ovens, refrigerator, large single tub sink, and lots of storage space. Also cupboard space for canned vegetables from our garden. (In the basement?) The kitchen should be semi open to the dining and living area but there should be some way to avoid looking at the mess (Eleanor's) when we sit down to eat, and a fool-proof kitchen fan to exhaust the aroma of Walter's famous garlic sauce. No disposal (we compost) and no trash compactor. We'd like maple butcher board tops and plenty of overhead light. The kitchen should receive direct morning sunlight and open into a

greenhouse or provide planting space within the kitchen.

Dining room

The dining room should probably separate the kitchen from the living room. Dining and living areas could be one big room. With some sort of implied or real divider? Steps? Built-in furniture? Change in ceiling height? View from the dining room is important. It's used constantly even though breakfast and lunch are usually eaten on the run. There should be space enough for a table which can seat ten or twelve in a pinch. Also room for a sideboard (antique) and extra chairs.

Living room

The living room needn't be large. We've given our piano to one of the children. We'd like a Franklin-type, airtight wood stove, bookshelves, a place for our record collection, turntable, and speakers, etc., and enough glass to capitalize on the good views on the south. We're partial to wide oak flooring—perhaps in the living room and dining room.

Bedroom wing

Our bedroom wing should be private, ample, and cozy. The bedroom might have a small wood stove in it—again, the Franklin type so that we can watch the open fire and then close it down. No rows of closet doors in the bedroom! Clothes storage should be all contained in one large walk-through closet, perhaps it could double as access to the bathroom. The bathroom must have natural light—an operable window—(next to the tub?). Could we grow plants in the bathroom? Maybe next to the tub. Tiled floor, nothing exotic. How about giving us a linen closet in here? What about a laundry chute? (Walter says no.) If eastern exposure is possible, we'd like that morning light instead of an alarm clock in the master bedroom.

Guest rooms

The guest rooms ought to be at the other end of the house—or on another level—from us. We love our grandchildren but we've spent enough late nights with our own children not to want to repeat those tribulations with another generation. The bedrooms can be moderately sized, each with a closet, and the bathroom should be accessible to the dinner guest as well as the overnighter.

Greenhouse

Walter's greenhouse needs southern exposure, should have outside access, and needs a small tool storage area adjacent to it. The floor must drain easily and we must make certain it can't freeze up. Could we use house heat to warm the greenhouse, and vice versa? Greenhouse size should be at least seven feet wide and twenty feet long. We're serious about plants!

Other needs

What else? The broom closet. Maybe in the mud room? The utility room? Could we add a root cellar? Will we need a furnace? Where

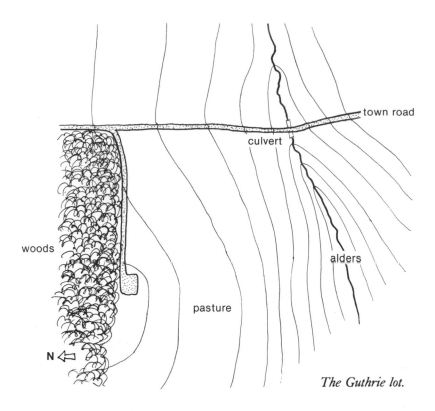

The Guthrie lot.

will we store firewood? We plan to burn a lot of wood. Do we need a basement? Oh, cars.

We want to spend as little as possible on the deification of the automobile. The BMW starts fine in cold weather, but we'd like to keep a roof over it. We may have a second vehicle for use around the place, an old pickup truck or a farm tractor. We've got lots of hand tools, a wheelbarrow, a roto-tiller, and Walter's table saw and drill press. We've also got a fifteen-foot Old Town canoe and boxes of junk we're not ready to part with. Cold storage for all the above is no problem.

Solar heated water?

Last-minute questions. Domestic hot water solar panels. Are they worth the installation cost? We've heard there's a 40 percent tax credit. What are the pros and cons of a composting toilet system? We don't want to have the garage doors central to the approach to the house. The garage needn't be attached but should be close by. What about electric lines to the house? Can we bury them? How much would it cost?

That's it for now. We'll see you on the land on Wednesday. We can't wait to get started!

126

Meeting on the Site

We met on the land just before noon on a warm, sunny day. The graveled town road leading to the property had not yet turned to mud. The late March thaw was yet to come. Snow cover had disappeared everywhere but in the darkest north slope recesses.

We turned west through an opening in a barbed wire fence and followed a frozen farm road a hundred yards up a gentle rise. To the right was a mature stand of sugar maples, a row of which edged the northern rim of the overgrown pasture we had entered. Looking southward we could see several successions of wooded knolls with a view of a distant peaked mountain off to the southeast.

The pasture sloped gently southward and showed no signs of ledge or protruding boulders. Winter-brown grasses covered the hillside, which was grazed every summer by a neighbor's Holsteins. I looked for juniper, ferns, or ground moss—telltale signs of ledge or water—and found none. At the foot of the hill, 500 feet away, was a dense thicket of white alders, a sure sign of water. To the east, I could see the end of a culvert poked out through the bare bushes where a brook crossed under the town road on its way west through the Guthries' alder bog. If there's still a half-inch pipe's worth of water running through that watershed in August, they've got a perfect site for a pond.

Ready for Work

Walter and Eleanor came prepared. They had a compass, topo maps, a transit, 100-foot tape, stakes, surveyors' ribbon, a shovel, and a five-foot iron bar. (There was also a full picnic basket and a bottle of California Chablis to celebrate the occasion.) For a couple of hours we pulled the tape this way and that, drove in stakes, pulled them out, and drove them in again. At noon we reckoned due south (*not* daylight saving time-south) by unanimous vote, confirming our simple observation with the compass. Eleanor worried that the compass pointed west of what appeared to be north. The approximately fifteen-degree variation between true north and magnetic

north in this area has confused many. For house-siting purposes, I find simple, thoughtful observation to be invulnerable to magnetic pull, surveyors' typos, or local myth: (the west branch of the river runs west).

South Slope

The transit showed us a one-in-six pitch, north to south, perfect for earth-covering one level. Before we arrived, Walter had driven an iron bar down four and more feet in a number of spots and found no sign of ledge. I suspected we'd find none at whatever depth we needed to go. I couldn't say exactly why except that there was no sign of ledge anywhere around. We were in a high alluvial valley. The eroded banks along the road on the way in showed nothing but varigated glacial till so typical to this part of New Hampshire. Walter needed reassurance that his site was ledge-free, but all I could do at the moment was to share my hunch, wait for a backhoed test pit, and explain that as a last resort, blasting is not prohibitively expensive anyway.

The site suggested several important pa-

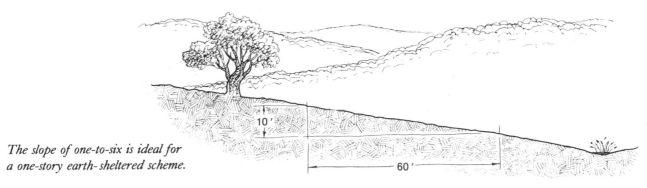

The slope of one-to-six is ideal for a one-story earth-sheltered scheme.

10′

60′

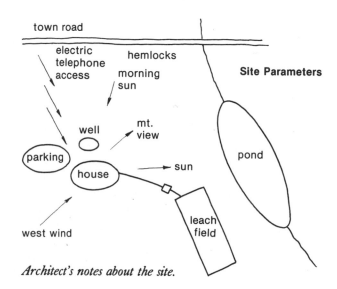

town road

electric
telephone
access

hemlocks

morning
sun

Site Parameters

well

mt.
view

parking

pond

house

sun

west wind

leach
field

Architect's notes about the site.

rameters to the design of the house. I sketched a site plan peppered with the following notes: Access from the east. View of peaked mountain to southeast. Morning sun available. How to downplay driveway and car parking? Hide driveway and town road with row of hemlocks? Southern view entirely open. Pond? Privacy at south or west of house. Electricity and telephone at road. Bring in underground. Leach field on flattened plateau to southwest of house. Drill well at northeast of house. Prevailing west wind. No views of interest to west (or north). Back of house near tree line. Park behind house between trees and house? Entry on east wall? East and south most important for views out of house and sunlight into house. Soil looks very well drained. Likely to be gravel. Order test pits dug to twelve feet, checking for ground water or ledge. Contact electric and telephone companies. Stake out driveway. Have percolation test done for state water pollution control board approval. Local permits? Building code?

Lunch with a View

We spread out the picnic on the hood of my car and swooned over the site, the view, the weather, and how brilliant we all were to be alive and well and eating Brie and roast beef sandwiches in New Hampshire in March. We speculated on the pond, the cost of the driveway ($6 to $10 a linear foot including a culvert at the curb cut), and a schedule for planting trees. Walter was deter-

mined to start digging. He'd already begun a compost heap the summer before, and proudly showed us the maples he'd pruned and the woodland grove he'd cleared of brush and deadwood.

The conversation about tree planting led to talk of our project schedule, and we laid out tentative dates for each stage. Within a month Geoffrey and I would present the Guthries with a design proposal. We expected, from past experience, to be able to solidify the design within six weeks after our first proposal and knock off the working drawings during the following six or eight weeks. That took us to the end of July. Letting the project out to bid would require a few weeks more. With any luck, we could begin digging by the end of August, five months from the first site visit. We toasted the peaked mountain. The sun sagged westward. Our fingers were numb with cold. It was time to go home, sit down, and draw.

The first doodle, one step toward blueprints.

The First Idea

Just as a walk around the block begins with a single step, a building is conceived with a seminal idea. The idea inevitably outgrows itself and evolves into other ideas. Eventually they coalesce to form a complex organism capable of performing a variety of functions. But there's always the "first idea." It may be replaced in importance by subsequent discoveries, but the beginning begins with *something,* no matter how grandiose or banal. I invariably begin with a rectangle.

For the Guthries' house, I put my trusted rectangle on the site plan near the trees at the top of the pasture, placing it near center of the ten-acre lot. The rectangle is a formality on my part, a habit. It provides a grid from which I can measure, a reference for structure and organization. The initial rectangle is

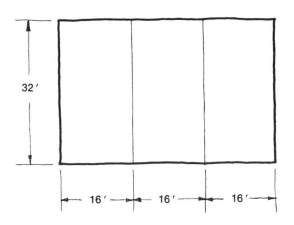

subdivided into sixteen-foot bays, the maximum practical framing limit for ordinary joist and rafter spans. Once this ritual doodle has been committed, I plug in that "first idea."

One Key Room

The kitchen belonged at the southeast corner; morning light into the kitchen, south light into the kitchen. The placement of the kitchen would ultimately dictate the placement of every other room in the house. I wasn't aware of that when I began, but it turned out to be unalterable, like a chess move that sets up consequences far beyond the immediate transaction.

Questions

An entry on the east wall gave access to the kitchen and living-dining areas. What about the two bedroom areas? How many levels would this house have? At this point, a second, more important consideration replaced the original kitchen-southeast corner notion. What kind of massing would this house—this site, this program—suggest? The one-in-six pitch on the hillside was perfect for an earth-sheltered, single-story configuration, but what about the interior zoning? I suspected the inherently linear nature of an elevational earth-sheltered scheme would require too much circulation space and reduce the three-realm concept to zoning-by-doors rather than zoning by specific, discrete areas. How, then, to maintain the planning requirements

without introducing additional and unnecessary buffer zones?

Answers

The answer was "vertically." One bedroom zone, earth-sheltered on all sides but the south, fills the lower "basement" level. Family zone at the middle entry level. Second (adults') bedroom area on the top floor. Three floors, three zones. The house will be cubelike with a gabled roof rather than the underground shoebox shape. Two-story greenhouse on the south side? Earth-shelter parts of the middle level as well? My pencil twitched all over the paper in a dozen sketches, plans, and cross sections. Piece by piece, the thing began to grow into a credible whole.

When it looked as though I almost had it, I turned off my drafting lamp and went for a walk. It's pure superstition, but I always do it. You don't abuse a string of good luck. You've got to know when to walk away (know when to run). The next day, you sneak up on the drawings and hope it wasn't all wishful thinking.

Why That Shape?

Partly due to the wisdom of the mitten vs. glove metaphor, (volume/surface ratio), partly due to its historic New England setting, and partly because of a stubborn reluctance to participate in contemporary architectural trends, I was drawn to the traditional,

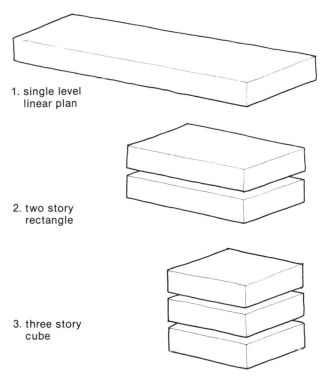

1. single level
   linear plan

2. two story
   rectangle

3. three story
   cube

Planning decisions.

compact, farmhouse cape configuration. Wide gable end to the sun, almost square in plan and cubelike in volume, it provided a staunch but versatile format within which a number of variations could occur. Water and snow from the pitched roofs would fall away from the greenhouse rather than upon it as in the typical double-envelope saltbox configuration. Glazing on the south wall of the upper floors would be vertical, rather than sloped, reducing the extent of summer overheating and winter heat loss. The north gable end inset provided handy but inconspicuous wood and tool storage and a "back door" entrance. The high roof pitch (12 in 12) would guarantee a good cooling chimney effect through the louvered vents and the potential for a dramatic high-ceilinged master bedroom. All the framing spans were under sixteen feet and there was nothing about the house that a competent builder of ordinary homes would be reluctant to undertake.

Over the next couple of weeks I pushed and shoved the house into shape. I committed myself to several basic premises and began calculations on how well they would perform. I called Marc, the solar engineer, and made an appointment to show him the progress prints. The house would combine the primary features of the earth-sheltered, double-envelope and superinsulated concepts. The site was perfect for full southern exposure, minimal north and west exposure, and moderate amounts of earth coupling, even with a multistory plan. Superinsulation could be added to any combination of the above. The

For storage of heat, figure on three to five gallons of water for each square foot of glazing. Thus, if your glazing measured ten by ten feet, for a total of 100 square feet, you would want to provide 300 to 500 gallons of water to store the heat.

question foremost in my mind was the design of the greenhouse.

If it went up more than one story, could it attain total solar self-sufficiency and provide some heat for the house as well? I guessed it would come close provided we included enough thermal storage mass (probably water-filled fifty-five-gallon drums under the benches). My work with earth-sheltered buildings rarely involved non-structural storage methods (rock beds, Trombe walls, and water tanks) because the heavy structural shells have enough inherent mass. Chimneys, masonry partition walls, and tile, brick, or slate floors generally preclude specialized storage systems. But with this greenhouse, where temperatures should never drop below 40° F. or rise above 75° F., I wasn't sure. I needed Marc's analysis to refine my simplistic design. I only knew we needed three to five gallons of water for every square foot of glazing, but there were further refinements that Marc would suggest to make the greenhouse perform optimally.

In his figuring, the key components would be:

1. Volume of heated space
2. Thermal resistance of greenhouse shell with various R-values assigned to each type of wall, roof, slab, window, door, etc.
3. Calculation of internal gain, if any
4. Calculation of solar gain
5. Calculation of infiltration, expressed in number of air changes per hour
6. Calculation of amount of mass required

to store solar gain and maintain the 40–75 °
F. range ideal for plants.

Push all of the above through the equations
including a few more assumptions such as:
figures for degree days, (7,870), extreme outside temperature for maximum △ T (−20 °
F.), soil temperature (assume 50 ° F.), space-
to-soil △ T (assume a cooling mode △ T of
− 25 ° F. and a heating mode △ T of + 10 °
F., and inside ambient temperature, 40–75 °
F. With these numbers Marc would miraculously test my seat-of-the-pants assumptions.
Then I would adjust according to aesthetics
and reality.

Marc's Report

Within a few days Marc's analysis was
complete and the hard cold facts were encouraging. His report read as follows:

Dear Don:

*Here is the preliminary solar analysis you re-
quested on the Guthrie house. The analysis of
this building is not simple, involving as it does
partial earth berming, a solar greenhouse, buff-
ering of 50° F. spaces, etc. To keep within the
time you allotted me, I've made educated estimates where appropriate, so remember that
these are ballpark figures.*

*Assuming a −20° F. design temperature, the
design load is 28,700 Btu/hr (8.4 kw). I've assumed entry/mud room temperatures of 50° F.,
greenhouse minimum temperature of 40° F.,
and below-grade temperatures of 50° F. I've*

*estimated infiltration at 0.33 air change per
hour (ACH), based on 0.1 ACH on the first
floor, 0.5 ACH on the second floor, and 0.4
ACH on the third floor. Changing these numbers affects both design load and annual auxiliary consumption considerably, and, since I don't
know all the construction details of the house, I
am wary of estimates here. I have assumed standard wooden doors as exterior doors. Foam-core
steel doors would save over two million Btu annually. This is a good energy investment but
many people find the steel doors unappealing to
look at. With these assumptions, the design load
breaks down percentagewise by section as follows:*

Section	% of Design Load
Windows	27
Exterior doors	8
Ceiling	6
below grade	5
Walls	20
Infiltration	34

*As you can see, once you insulate well, it's all
going out the windows and in air change. It
would be even more lopsided if the glass on the
first floor were not buffered by the greenhouse,
or if the house was constructed with 0.75 ACH.*

*I used Burlington, Vermont, solar and
weather data for the calculation of annual auxiliary energy load. My assumptions include:*

- *No heat provided to the greenhouse or
  storage spaces*
- *500 kwh/month utility contribution*
- *Four occupants*

The gross annual heating load is fifty-nine million Btu. Of this, about seventeen million Btu are supplied by direct gain passive solar heating through the second and third floor glazing. Between six and eleven million additional Btu may be supplied by the greenhouse, depending on greenhouse construction, temperature limits, heat extraction methods (e.g. a blower would raise the yield), and the ability of the house to absorb the extra energy from the greenhouse at a time when it is also storing direct gain energy. If you are interested, supply me with construction sections and an interior finish schedule and we can determine if adequate thermal mass is present.

Therefore, total solar contribution is between twenty-three and twenty-eight million Btu an-

nually. Thirteen million Btu are supplied by the utility usage and the occupants (intrinsic energy sources), and that leaves the auxiliary consumption somewhere in the range of eighteen to twenty-three million Btu annually. This represents 1¼ to 1½ cords of wood, 180 to 230 gallons of oil, or 5,300 to 6,800 kwh per year in auxiliary fuel consumption.

I have several suggestions of ways to improve this design. Please bear in mind that I don't know your clients or their specific program, so some of this may be inapplicable.

Given the design at present, consider the following:

1. An airlock entry on the westward living/dining room door will reduce air infiltration to these spaces substantially.

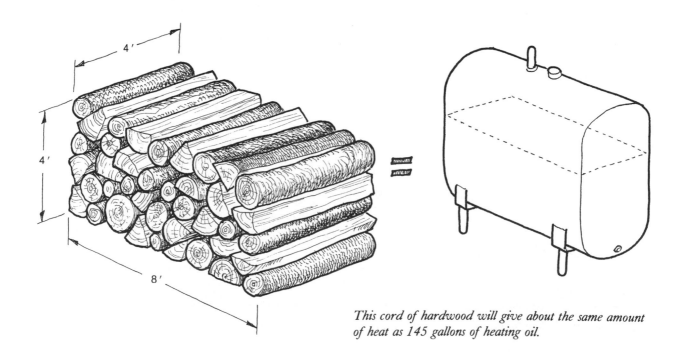

This cord of hardwood will give about the same amount of heat as 145 gallons of heating oil.

2. I am not familiar with the construction details you intend to use, but accepted superinsulated practices such as no recessed ceiling fixtures, no holes in the vapor barrier for electrical receptacles, foamed-in-place windows and doors, a flexible section on the (carefully caulked) plumbing stack(s), etc., should be followed. Also, exhaust fans should preferably be ducted down and then out, to prevent the ducts from acting as chimneys, removing heated air from the house. Should the house be truly supertight, an air-to-air heat exchanger might be appropriate.

3. Outdoor air, via closable ducts, should be supplied at or near the inlets to the wood stoves. (This is to prevent cold drafts created by the stove drawing its combustion air from cracks around doors, etc.)

4. I recommend small stoves that can run efficiently at low burn rates. I have had success with the Jotul 602 in my own superinsulated home.

5. Finally, the master bedroom is somewhat overglazed, at least to the point where 15–20° F. winter clear-day temperature swings are likely. There are no overhangs due to the building and window geometry, making summer overheating a likely possibility. Therefore, I suggest that the glass area be reduced to no more than fifty square feet net (about two-thirds of present amount). As you know, nighttime window insulation will improve solar performance and comfort, but is expensive and needs to be operated to be effective!

I hope the information and suggestions here

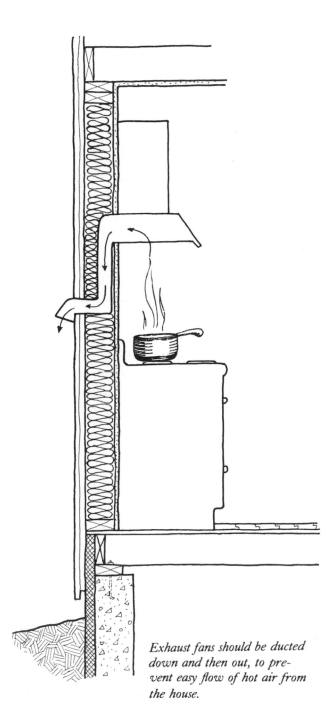

Exhaust fans should be ducted down and then out, to prevent easy flow of hot air from the house.

are helpful. If you would like to pursue the engineering of any of these changes, we would be delighted to be involved.

Yours truly,

Marc Rosenbaum
Energysmiths, Inc.

Airlock Costs

Marc's analysis surprised me in one minor respect. Items #1 through 4 were solid, conservative reminders of the small but important details that make a building perform optimally. I put them in the category reserved for minor-changes-subject-to-aesthetic approval.

Airlock Rejected

An airlock entry on the west living/dining room door would reduce infiltration, to be sure, but construction costs and the architectural awkwardness of a vestibule on that wall would far exceed the savings in energy costs. The airlock would not be added. Marc's comments on construction details, especially in regard to vapor barriers and down-flow exhaust ducts, were to be taken seriously. They would be automatically included in the Guthries' house plans and specifications.

Outside air for the woodstoves would be furnished via four-inch insulated PVC pipes run between floor joists from the outside wall closest to each stove.

As for types of airtight stoves, I would add to Marc's preference for Jøtul the equally efficient Lange, Riteway, and Vermont Castings models.

Foam-cored, steel-clad doors, especially the Brosco "Ever-Strait" line, are close in quality to a solid wood door and offer far superior insulation and built-in weatherstripping. I would recommend that we use them on all the Guthries' exterior doors.

Expected More Heat

Item 5 surprised me. I was disappointed to find that so much south-oriented glazing in the bedroom did not contribute more usable heat. I've always considered double-glazed, south-pointing windows to be net gainers, but the recent Department of Energy (Balcomb) analysis stresses that an overglazed, undermassed building can't absorb the direct solar gain it receives. More heat than can comfortably be lived with will probably be gained, then dumped during ventilation, unless substantial amounts of mass are available for storage.

Another layer (or two) of Sheetrock is an inexpensive option for adding mass. We would consider that for the master bedroom.

Overglazed or not, all the Guthries' windows and skylights would be triple-glazed except for the sloping greenhouse glass, which would be tempered, double-glazed, sliding glass door replacement units.

As far as performance was concerned, the Guthries had no complaints. Their space heating costs for an entire year would be under $300 a year with oil, or "free" if Walter cut his own wood. At a moderate pace, he could cut, split, and stack his winter's supply (one and one-half cords) in two days.

More importantly, the house would be a delight to live in and simple to maintain. The same-sized house, twenty years ago, would have consumed four or five times the energy for space heating. And yet the Guthries' house hardly qualifies as a miracle. Its simple, low-tech, easily assembled parts and its plain vanilla style make a convincing case for a new standard of housing that can dramatically reduce the amounts of energy consumed by space heating. Furthermore, once the techniques have been demystified—which they have quickly become—we can reinvest in architectural expression and enjoy the most important part of our sheltered environment, the spirit and vitality of space, form, and light.

An Important Meeting

I called the Guthries and told them we were ready to present our solution. Floor plans, elevations, and cross sections were tacked up on the office wall. A copy of Marc's report framed one side of the drawings. The ball was soon to be in their court.

We were all a little nervous when we met in my office the next day. The drawings looked good to me, but I couldn't help but

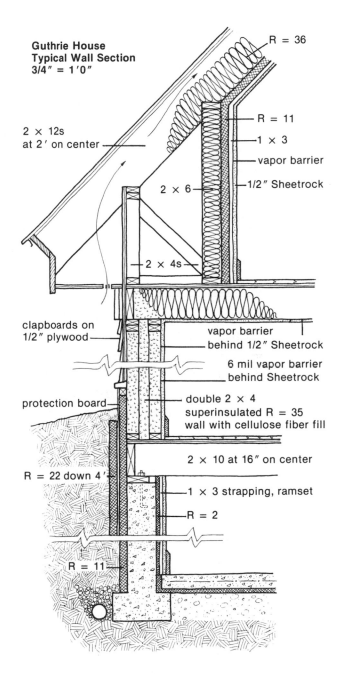

**Guthrie House
Typical Wall Section
3/4″ = 1′0″**

R = 36

2 × 12s
at 2′ on center

R = 11

1 × 3

vapor barrier

1/2″ Sheetrock

2 × 6

2 × 4s

clapboards on
1/2″ plywood

vapor barrier
behind 1/2″ Sheetrock

6 mil vapor barrier
behind Sheetrock

protection board

double 2 × 4
superinsulated R = 35
wall with cellulose fiber fill

2 × 10 at 16″ on center

1 × 3 strapping, ramset

R = 22 down 4′

R = 2

R = 11

Wall section of Guthrie house.

South elevation

wonder if I had misjudged the Guthries' intentions, if I had dared too little with the simple cape massing, if the floor plan was too mundane. It was, after all, not a flashy house. But then, I consoled myself, we had dismissed flashiness as an unworthy aspiration.

I was prepared to explain that the house was in no way profound. It would certainly claim no place in the evolution of architectural ideology. And yet, it worked extremely well. It worked in all the ways a house for this time and place should work, and if the spaces were not positively stunning, neither were they ordinary. From the huge window in the master bedroom to the kitchen-dining-living-room to the lower-level bedrooms and plant-filled greenhouse, there was an ongoing variety and vitality that seemed appropriate and inviting. And at the bottom line, I knew I could tell them they could heat the place all winter long with a cord and a half of wood.

138

Nervous Too

As I learned later, they were nervous, too, but for reasons quite different from mine. Walter confessed that he worried I would be too creative for his pipe-smoking metabolism and pull an "ego trip." He was visibly relieved when he saw the gabled roof and the square plan. He was like a kid with a new toy when he saw the greenhouse. Eleanor referred to it as a sunspace. To Walter, it will always be a greenhouse. But it wasn't connected to the kitchen!

Issue of Greenhouse

The greenhouse quickly became an issue. Greenhouses were supposed to be all glass. This one was half buried, and the only glass was on the roof. Wouldn't it be too hot in

North elevation

summer? I explained that they would have to shade the glass in summer. Commercial greenhouses—with sloped roofs—are usually covered with a coat of whitewash during the summer. By the time the rain has washed it off, it's autumn. The sloped glass roof lets in more light than is needed during the summer, but is ideal in the spring when most greenhouse planting is done. During the winter months in New Hampshire when the sun sinks as low as 22° at noon, vertical glazing deflects approximately the same amount of sunlight as a 45° pitch. But vertical glazing reflects too much valuable sunshine during those critical springtime planting months. Heat lost through vertical glazing is admittedly less than the heat lost through the same area on a 45° plane, but plants respond best to light from above.

The greenhouse ventilation was designed to work either mechanically or by natural con-

vection. In summer, the open door brings in air at grade. As the air is warmed, it rises and escapes through the trap doors in the sloping roof above the glazing. (The trap doors are opened from the greenhouse with a thermostat-activated solenoid.) Shading will be necessary for the summer months, due to the slope of the glass. Bamboo or synthetic fabric shades can reduce sunlight infiltration by 50 percent and water storage tanks will help to moderate extreme temperatures, and the overhang at the second-floor ceiling will shade almost half the kitchen-dining room glazing in mid-summer when the sun is almost 80° above the horizon at noon.

During the winter months, excess warm air from the greenhouse can be let into the house through trap doors just above the floor in the dining room.

Walter and Eleanor are sold on the greenhouse, but Walter wants his herbs next to the

139

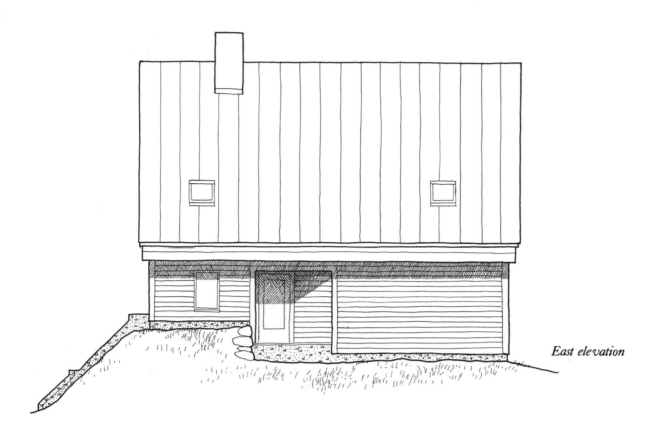

East elevation

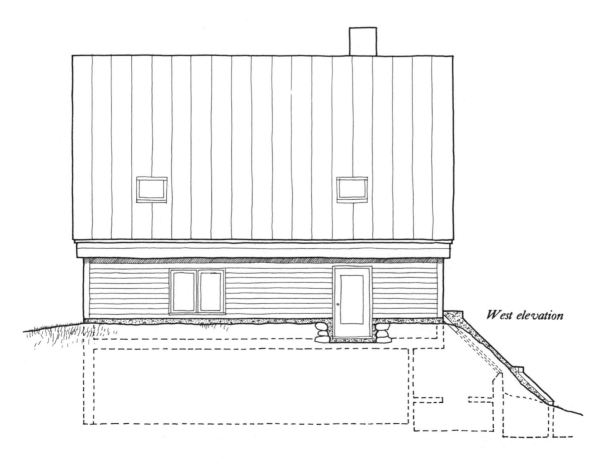

West elevation

kitchen. I point to the planting tray between the kitchen sink and the windows. It is not a greenhouse, but it works. The label isn't important. Walter grouses about the "planter." It's *not* a greenhouse. It's a copper-lined pan, ten inches wide, eight inches deep, and fifteen feet long. We humor him. As a growing space for herbs it works. And it can be watered with the sprayer at the kitchen sink. The *real* greenhouse below makes up for the disappointment. He adds up its strong points—earth-tempered on two sides, backed up to the bedrooms on another, and glazed at the optimum angle. He likes the tool storage area in the airlock entry and finally lets go his kitchen-greenhouse dream. He likes the sunken jungle room full of blooming, creeping, foliating flora. He'll grow his herbs by the kitchen window and reserve the serious work for down under.

Bedroom Window

Eleanor asks about that big master bedroom window. She's worried about how much it's going to overheat the room. She hasn't yet seen Marc's report but she's already identified the problem area. I'm in good company. We discuss adding Sheetrock for added mass, but it appears the window has appeal. It looks huge, we all agree, and it can't be inexpensive. I've done my homework on the window costs. It would come to about $1,800. We talked about cost effectiveness. How do we evaluate the "payback" from an investment which delivers delight rather than dollars?

Looking over the house as a whole, I plead the case as a permissible, minor extravagance, hardly an act of fanciful indulgence. I sketch an interior perspective of the master bedroom with that huge opening at the edge, like a Jules Verne spaceship cockpit, or, Eleanor quips, a poor man's Palladio gone square. As they visualize it more thoroughly, they like it, even though it may make the room too hot during the summer months. The window will remain.

The Bathrooms

There is a point, in all forms of negotiation, at which each party must review the remaining agenda and decide which issues would be willingly sacrificed in the interest of a common goal. After the greenhouse and bedroom window disputes were credited to my account, I decided to bring up an area which I had solved badly—the distribution of the bathrooms. With luck, the Guthries would have a hand in correcting the problem and share in the evolution of their house design.

I'm convinced that part of a residential architect's responsibility is to encourage the client's involvement in the design process, and admit some (small) degree of mortal incompetence. In the back of every client's mind lurks the potential for an attitude which sounds like "... after all, *we're* the ones who have to *live* here. ..." Nothing could be more true, and yet a trusting and responsible dialogue throughout the design process will eliminate the likelihood of this kind of dead-ended response and keep the spirit of collective enterprise alive.

There is no bathroom on the entry level, and the master bath is inaccessible except through the master bedroom. This means that the dinner party guest or the extra overnight guest who may be put to bed in the study will use the "basement" bathroom. Walter thinks

nothing of it. Eleanor is skeptical but admits she likes the privacy of the master bedroom suite and doesn't want to sacrifice any of the entry-level footage for even a half bath. We agree that the logical location for an additional bath would be off the entry, borrowing space from the mud room. Maybe the washer-dryer could be moved to the basement. Walter insists we forget it, citing extra

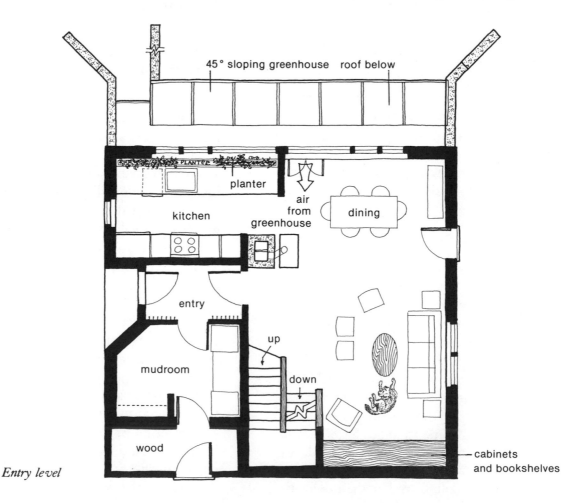

Entry level

cost, limited space, and a brief review of the relationship between exercise and longevity. So much for the lost bathroom.

The Kitchen

The kitchen seems to work perfectly as far as both cooks are concerned. We agree to polished slabs of slate and butcher board for the countertops, a pot rack hanging from the ceiling over the range, and oak cabinets. The window to the east and two windows to the south are operable. The view from the kitchen sink, to the hills beyond, is a topic of animated enthusiasm. The cook-to-dinner guest proximity seems workable enough, and I suggest that the inevitable after-dinner mess is best obscured from the living room with a light switch. The wood stove location is ideally centralized, although wood storage is not as close to the stove as it might be. We discuss the possibility of a wood box under the west end of the southside kitchen counter.

The entry and mud room will be minimally heated—to 50° F.—just enough in the mud room to make washer-dryer use comfortable. The west wall of the mud room and the south wall of the entry (as well as the outside walls) will be insulated. The heating load on the entry level will include only about 60 percent of the 960 square feet gross area.

The lower level is so straightforward as to preclude much comment. The bedrooms will be heated only when occupied, either electrically or with six-inch ducts with in-line fans drawing heat down registers from the floor above. The bathroom will require no more than a radiant ceiling fixture. The storage room and root cellar will remain an even 50 ± ° F. year round and the greenhouse will heat itself. Three out of four walls on the lower level are earth-coupled. The fourth is tempered by the greenhouse. The colonial-era housewright's rule of thumb held that if seven-eighths of your basement was below grade, your potatoes would never freeze. This assumed a house above the basement which might have a fireplace or stove in more than one room, but would, in effect, be only marginally heated. The Guthries' lower level is not only more than seven-eighths below grade, it is insulated on all below-grade walls, covered with a "warm" top, and most importantly, receives considerable solar radiation from the south walls of the two rooms that require heating. Because the lower level is an occasional-use area, the heating requirements will impose a minimal impact on the overall performance of the house.

Master Bedroom

We move on to the upper-level master bedroom suite and study. Although the study will turn out to have the greatest Btu per square foot heat loss of any rooms in the house, its requirements are still miniscule. Its only window faces north, its skylight is at a 45° angle, and two walls and the ceiling are outside surfaces. But its maximum heat loss is still only 2,100 Btu per hour. At 3.4 Btu per watt, the room could be heated with six 100-watt light bulbs and a canary. This makes even the "coldest" room not so difficult to heat after all. The master bedroom has three outside walls and ceiling, but its immense south-wall glazing provides more than enough solar radiation to offset the losses through the heavily insulated shell. As Marc pointed out,

the room suffers from too little storage mass. We decide to add double or triple layers of Sheetrock to increase the room's thermal capacity. The east and west walls in the bedroom are four-foot high "knee walls" which make the room appear, in plan, wider than it effectively is. It is still an unusually large room and invites subdivision into two areas. The bed will have its headboard against the

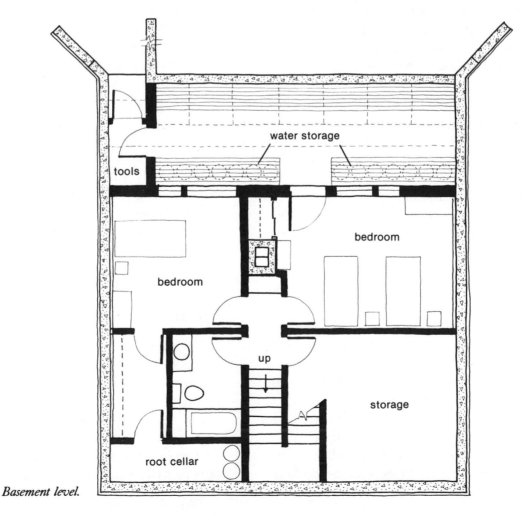

Basement level.

north wall to the right of the door. The area near the dressing room will become a sitting/reading alcove. The wood stove is nicely balanced between the two zones.

There will be no heating difficulty on this upper floor for a number of reasons. First of all, the stairwell acts as a chimney for warm air generated by wood stove or greenhouse collection on the floors below. With the study or master bedroom door open, heated air from below will maintain the rooms at a temperature close to the entry-level ambience. Even with the doors shut, the warm air will stack up in the hallway and simply temper the surrounding walls. Secondly, since the entry level is heated more than any other part of the house, the uninsulated floor between the entry and upper level will conduct heat upward. (Needless to say, the entry and mud room ceilings and floors will be insulated against the warmer parts of the house. The same goes for the wood storage and tool storage rooms.) Solar input through the great bedroom window exceeds the heating needs during the sunny daylight hours, and the smallest wood-burning stove will provide ample warmth on the coldest February night.

Sold

The overall layout seemed to work, and the Guthries liked what they saw.

The house would be sited at the high point of the knoll, fifty feet to the south of the tree line. To the east of it, with its gable-end doors at the edge of the trees, would be a twenty-four-foot square garage with a lean-to shed off to the rear.

Cars in the parking area between the garage and the house would not be visible from inside the house, and the garage would be hidden from view by the huge maples along the edge of the forest.

A private outdoor sitting area would be developed on the southwest side of the house, with view to the mountains and the valley below.

A Hybrid House

Generically, the house is a hybrid. Its superinsulated walls and roof are its foremost energy-conserving feature.

The earth-sheltered lower floor and greenhouse utilize the sloping site in the most economical way possible. The sub-soil walls need no exotic (expensive) waterproofing and the structural demands on the reinforced concrete walls are only slightly in excess of a standard basement wall.

The greenhouse and the large expanse of south-oriented glazing borrow the most practical feature of the double-envelope house—the passive solar glazing and the sunspace/collector—and make a major contribution without the expense and redundancy of the double-shell convective loop.

This combination of systems allows the appropriate use of the appropriate tools in a *selective* way rather than the full-bore dogma that so often accompanies the systemic approach.

From the Guthries' personal standpoint, this house would allow them to pursue the kind of life they intended to enjoy in their later years. There was little imposition of "style" except for an implicit informality and a selective borrowing of regional New England building forms. Their furnishings and habits would fit the interiors nicely. The gabled roof and bleached cedar clapboards would make their new home an instantly familiar and comforting part of that worn, heroic, New Hampshire hill country. Theirs is not a house for the freeway through Malibu. It is not a house for mirrored walls and chromed chairs with European nametags, nor is it a house for blown-dry hair and fast food, fast talk, fast-lane sensibilities.

Working with Nature

It is a house designed to work with nature, not against it. The four seasons lend as much to the decor as the furniture itself. Firewood stored in September provides heat all winter and ashes for the garden in May. Snow and mud-clad boots will be no strangers to the tiled entry hall floor. Summer breezes blow through the open screened windows, bringing hints of hayfields, rain a-coming, and the hot, lazy smell of dust. The low winter sun floods

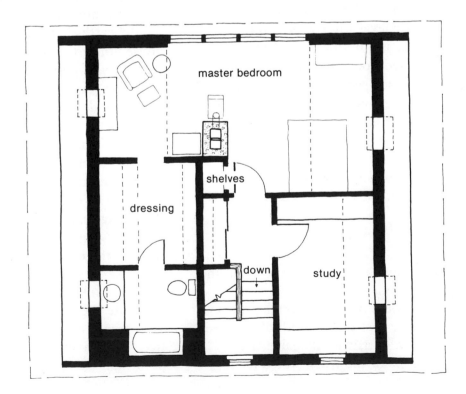

Second floor level

the house with light and warmth and ever-shortening shadows as the days lengthen towards the vernal equinox. Greenhouse aromas, humidity, and the ongoing cycle of seedling-to-flower provide a living journal of nature's awesome persistence. Rather than overwhelm its environment, the house participates in harmony with the elemental forces that give life to that particular place. And the extent to which it succeeds will depend upon the degree to which those fundamental cycles of heat and cold and moisture have been appreciated and accepted. Nature bats last, and when it comes time to choose up teams, there's no question as to which side we hope to be on.

A friend of mine, an avid sailor and consummate pragmatist, tells a story about a latter-day Barnacle Bill, a crusty old salt who was born to the sea. Every autumn, the intrepid old Ahab sails his rusty tub from Cape Cod to Florida. In the spring, he sails back north to preside over the boatyard's summer trade, contemptuous of the fortunes landlubbers spend on esoteric fittings and navigational devices. One day, during a dialogue on exotic ship-to-shore radio equipment, one of the big shots asked Bill how he navigated his way to Florida and back every year. The old man squinted at a sea gull and took another puff on his pipe. "It's no problem a'tall. When I head down to St. Petersburg, the United States is on the *right*. When I sail up north again in the spring-time, it's on the *left*."

If this book has served its purpose, it should have been consistent with the kind of observations that Barnacle Bill would approve of. The Solar Age has proposed exciting, challenging, and essential changes to the ways in which we interact with our natural resources and environment. Our computers and high-tech industries may provide easy access to the new, appropriate technologies, but in the end we must count on our common sense and basic physical evidence—and remember that no matter what it's called, smoke goes up, water goes down.

INDEX